# THE COMPLEX LIVES OF BRITISH FRESHWATER FISHES

## Praise for the book

This book is an excellent read and opens your eyes to the secretive world of freshwater fish. Even as someone who has been involved with fish and fisheries for all of their professional life, I still found out lots of things I didn't know. I particularly liked the chapter on what have freshwater fishes ever done for us, and who doesn't like to dig a bit deeper into the weird and wonderful sex lives of fish? This book would make a great addition to the bookshelves of anyone who works with, or is interested in, fish and their biology. You cannot fail to pick it up and learn something new, even if you think you know all there is to know about fish.

*Paul Coulson, MIFM MIfL, Director of Operations, Institute of Fisheries Management*

Mark's mastery of 'everyman' science and Jack's outstanding wildlife photography make this a game changing reference book, bridging the gap between anglers and fishery scientists and opening the full depth and breadth of our fish fauna up to the public. It is an outstanding work, and one to which I shall regularly refer.

*Ian Welch, Fishery scientist and angling writer*

When Jack told me last year that he was working on this with Mark my first thought was, 'Wow! A combination between Jack and Mark? I must buy that book when it comes out.' My second thought was, 'Oh, I hope it is not going to be another technical book on freshwater fish – I have loads of those.' Thankfully my first impulse was the right one. Yes, there is loads on life cycles, feeding habits and sex life of freshwater fish, but described in a very engaging manner – you can tell they are both anglers. Plus there are some fascinating facts and details: did you know that a salmon is a closer relative to a camel than a hagfish? Did you know that during the war they canned perch in Lancaster, adding Yorkshire relish and tomatoes (didn't catch on)? Did you know that in 1808 sticklebacks were so plentiful in the Fens that they were netted and used as a fertilizer on the fields? Did you know that in America they use bluegill sunfish to protect against terrorist acts? If you didn't, then buy the book and read it!

*Mark Owen, Head of Freshwater, Angling Trust*

An accessible and comprehensive guide to the fascinating lives of British freshwater fishes.

*Matthew Ford, IUCN Freshwater Fish Specialist Group*

For those who enjoy the eel, the chapter suitably headed 'Curious and Curiouser' is illuminating with many new stories. Highly recommended, especially and fittingly as eel numbers are slowly increasing.

*Andrew Kerr, Chairman, Sustainable Eel Group (SEG)*

# THE COMPLEX LIVES OF BRITISH FRESHWATER FISHES

**Dr MARK EVERARD**

*with photography by*
**Jack Perks**

CRC Press
Taylor & Francis Group
Boca Raton London New York

CRC Press is an imprint of the
Taylor & Francis Group, an **informa** business

First edition published 2020
by CRC Press
6000 Broken Sound Parkway NW, Suite 300, Boca Raton, FL 33487-2742

and by CRC Press
2 Park Square, Milton Park, Abingdon, Oxon, OX14 4RN

© 2020 by Taylor & Francis Group, LLC

CRC Press is an imprint of Taylor & Francis Group, LLC

ISBN: 978-0-367-44032-9 (hbk)
ISBN: 978-1-003-00760-9 (ebk)

**Visit the Taylor & Francis Web site at
http://www.taylorandfrancis.com**

**and the CRC Press Web site at
http://www.crcpress.com**

# Contents

# *Author*

**Dr Mark Everard** has spent his life connected with water, fishes and aquatic science. He has been active as a researcher, policy advocate, angler, author and broadcaster in Britain and Europe, but also across the developing world where his work includes sustainable water management for linked human security and ecological benefits.

Mark is Associate Professor of Ecosystem Services at UWE (the University of the West of England, Bristol). Mark's many other roles include Ambassador of the Angling Trust, science advisor to the Salmon & Trout Conservation UK (S&TCUK), member of the Conservation Policy Committee of the Wiltshire Wildlife Trust, science advisor to the Mahseer Trust, member of the Science and Technical Review Panel of the international Ramsar Convention, and life member and former Trustee/Director of the Freshwater Biological Association. Mark was instrumental in the founding of BART, the Bristol Avon Rivers Trust, which he still supports as part of a long-running commitment to Britain's thriving Rivers Trust movement.

Mark is also a frequent broadcaster on radio and television, gives many presentations to interest groups and is the author of many books, scientific papers and magazine articles, many of them addressing fish ecology and fishing. Mark is a passionate angler, getting out whenever he can after coarse, game and sea fish, with an enviable track record of specimen fish catches and a particular passion for roach and the rivers that support them.

In the angling press, Mark is often referred to as 'Dr Redfin' for his passion for the roach!

For more on Mark and his work, see www.markeverard.co.uk.

**Dr Mark Everard holding a grayling. (Image © Dr Mark Everard.)**

# Photographer

Based in Nottingham, Jack Perks is primarily a freelance underwater and wildlife cameraman, having worked on various BBC nature shows like *Springwatch*, *Countryfile* and *The One Show*. He's also filmed for angling series: *Mr Crabtree Goes Fishing*; *Fishing TV*; and *The Pursuit*. Jack has been awarded in both British Wildlife Photography Awards and Underwater Photographer of the Year.

A passionate angler and naturalist, Jack tries to use his work to highlight fish, having filmed every species of freshwater fish in the UK and organising a vote for a national fish (the brown trout won!).

Jack has written two books: *Freshwater Fishes of Britain* and *Field Guide to Pond and River Wildlife of Britain and Europe*. He has also written for many magazines including *BBC Wildlife*, *Outdoor Photography* and *Diver*, with regular columns in *Fallons Angler* and *Angler's Mail*. Jack is sponsored and works with Affinity, Hammond Drysuits and Fotospeed.

He's passionate about teaching others and, as well as talks and private photo workshops, he runs short courses on wildlife photography with Nottingham Trent University and is a part-time lecturer for the MSc in Biological Photography and Imaging course at the University of Nottingham.

Jack is also a trustee of the Trent Rivers Trust.

For more on Jack and his work, see www.jackperksphotography.com.

**Jack Perks holding a roach. (Image © Dr Mark Everard.)**

# INTRODUCTION

There are many books on 'freshwater fishes', an internet search yielding literally thousands. Many of these cover fishes of the world, with a fair smattering addressing species encountered in the British Isles and Europe. Some, including two that I have written myself, cover all species encountered in Britain. So, if you are after detailed listing of species with their key features and other technical aspects, the literature already serves you well. Equally, many books also cover angling for these fishes, again including a few of my own.

This, however, is not one of these kinds of books. It is instead principally about the fascinating and complex life histories of Britain's freshwater fishes, the many things they do for us (ranging from food, ornamentation, sport and cultural identity) and their importance for conservation as part of the living ecosystems upon which we all depend. Latin names are used sparingly and mainly only on first acquaintance with each species. Technical terms are also used only where necessary, but are also explained. (The Glossary at Appendix 1 will help you navigate terms that may be unfamiliar.)

For many people, the eye bounces off the surface film of Britain's diverse still, flowing, fresh and brackish waters, the mysteries lying below remaining substantially out of sight and so out of mind. This book ushers you at least a little way beneath the surface film and into the wonderful and fascinating world of the fish species encountered there, and their interactions within living ecosystems and our lives.

Too often, fish are regarded separately from other elements of wildlife. This is a false dichotomy. It is also to the substantial disadvantage of the ways we manage the ecosystems of which they are significant functional elements, as well as prejudicial to humanity in terms of the many benefits that these ecosystems confer upon us.

Fish are fascinating, uniquely so but also in these wider ecologically and culturally connected contexts. I hope that this book will open your eyes at least a little to Britain's freshwater fishy wealth and the many ways in which it enriches our lives.

As indicated by its title, the focus of this book is on British freshwater fishes. But many species found here also occur across the rest of Europe, though, owing to biogeographical history, the continental mainland has a far richer freshwater fish fauna than that of our small islands. Our historic connections with North America also result in a number of common species. Furthermore, our habit of moving species between global regions, often with the best of intentions yet not uncommonly with adverse consequences, also results in the species appearing in these pages being now far more widespread globally. But what is of generic relevance, not merely across Europe and North America but globally, is the detailed consideration in Chapter 3 of 'What have freshwater fishes ever done for us?' The chapter addresses the many benefits, often now referred to as 'ecosystem services', that fish and fishery ecosystems confer upon humanity.

By whatever happenchance or cause, water has been my lifelong fascination. Fresh waters pervade my earliest memories, education (a first-class honours degree and PhD in aquatic sciences), professional life as both a scientist and in international development, in my various writings, choice of home always by rivers and avid interest in angling and pet fish. I'm even fishing a river in the single movie in which I have acted, and one of my albums of original music is all about water. Water is central to the work I do, through its multiple roles as a fundamental resource underpinning, or impeding, a sustainable future.

Fish have always been integral to my life throughout my conscious history. As an avid nature watcher, as an angler, as a scientist and as a keeper of pet fish, fish have always swum in both my subconscious and conscious minds. Some might say that being born under the astrological sign of Pisces is contributory, whilst others may dispute the influence of far-off galaxies on such daily predilections. Whatever the reason, fish, and particularly the fishes of fresh water, have loomed large throughout my life, not to mention my culinary and recreational preferences.

Memories from the age of two, or thereabouts, are fresh in my mind of frequent visits to a farm dewpond close to my then Kentish home. Tiny though the pond was, though in my mind at the time a vast and mysterious expanse, it teemed with three-spined sticklebacks (*Gasterosteus aculeatus*), feisty tiddlers that we will meet many times throughout these pages. Armed with black cotton, a few worms and a rustic rod snapped from a nearby hedgerow, this was the stuff of many a piscatorial drama! The unfortunate worms were tied amidships on the end of the cotton, with no hook necessary. Lowered into the dark waters, it was never long before a stickleback gorged itself on one end of the worm and could be swung to land for inspection, admiration and subsequent release. But, let me assure you, this is no book about the art of angling. I've written plenty of those if that's your interest. Rather, these experiences have contrived to shape a life in thrall to the fishes of five continents, in all of which I have subsequently worked and also observed and caught fish.

As we've already observed, this is not so for everyone. The gaze of a surprisingly large proportion of the general public often fails to penetrate the surface film of the diverse waters in which fish swim. But Britain's freshwater fishes influence our lives in many, often unsuspected ways. I hope that this tome will shed light on some of the diverse ways in which they do.

# WHAT IS A FISH?

Before we crack on into detail, let's pause a moment to pose the deceptively simple question: *What is a fish?*

## 2.1 THERE IS NO SUCH THING AS A FISH

We all know what fish are, don't we? After all, fish are familiar enough in the windows of fishmongers, on our plates, in aquaria or garden ponds, as targets for our fishing interests, or else seen wild in our fresh waters. Some 32,000 species of limbless vertebrates found in aquatic habitats around the world are classified by science generally as 'fish'. However, after a lifetime's study of fish and other creatures, the American evolutionary biologist Stephen Jay Gould came to the stultifying conclusion that *there is no such thing as a fish*.

Gould's line of thought was, with hindsight, blisteringly clear. If you live in water, you need some basic apparatus to survive. Fins or fin-like appendages are the best way to propel a streamlined body through a viscous fluid, gills or gill-like structures help exchange gases and so on. Gould reasoned that the aquatic creatures, most of them not closely related to each other, started their evolutionary lines in the sea and, regardless of ancestry, evolved similar fish-like features as adaptations to survive underwater. A salmon, for example, is a closer relative to a camel than it is to a hagfish.* As life on land is a more recent evolutionary occurrence, this huge genetic diversity is a lesser feature of terrestrial animals. Gould's point is that as the product of 500 million years of evolution, 'fish' is something of an artificial category covering a wide array of aquatic creatures sharing some similar attributes.

So, if there is no such thing as a fish, perhaps this book should end at this point! Mercifully, we can define what we mean by 'fish' a little more tightly in the context of British fresh waters to include the teleost (the bony fishes) as well as some non-bony members (the lampreys). This assemblage of freshwater fishes is fascinating in its diversity, also constituting a too-often-neglected element of our natural heritage.

## 2.2 FISH AND OTHER WILDLIFE

Fishes have enjoyed a disconnected relationship in public perception, as also in legal definition, with the rest of the animal world. So why have fish so often formerly been a relatively neglected member of the fauna or been seen as something apart from the rest of 'wildlife'?

In part, as I elaborate later, this is because the eyes of many, perhaps most, of us terrestrial beings fail to penetrate below the water's surface to appreciate the wonders of the aquatic realm. But also, where recognised, fish are often not considered 'real wildlife'. The situation

---

* Campbell, A. and Dawes, J. eds. (2005). *Fish, What is a?* Encyclopedia of Underwater Life: Aquatic Invertebrates and Fishes. Oxford University Press.

is stark in India, the region in which a lot of my work is undertaken, where the Indian Wildlife (Protection) Act 1972 (and subsequent amendments) prohibits the killing of animals and plants across all Indian states (except Kashmir and Jammu, which has its own Wildlife Act, at least up to reforms instigated in 2019). The Act includes six Schedules relating variously to total protection with stringent penalties for infringement through to limitations on hunting, cultivation and other uses. However, the definition of 'animal' under the act includes amphibians, birds, mammals and reptiles. Fish are conspicuous by their exclusion from this definition, unlike other groups of wetland animals, with unfortunate consequences for their conservation. Fish, under strict legal definition, are regarded principally as food rather than wildlife.

Whilst we can observe such disconnections in perceptions of fauna when looking overseas, the situation in Britain is no less fractured, where it is common for 'fisheries' to be managed in separate organisational departments than 'conservation'. Here, it is primarily the recreational or food values of fish that set them apart from the rest of nature in the eyes of the public and policy makers. In reality, fishes are indivisible and fully interdependent elements of natural systems. Take fish away and the functioning of the ecosystem changes drastically. In some nature reserves, for example, the Cotswold Water Park in Gloucestershire, there is positive management for fish, but this is mainly to ensure that there is an adequate food resource to meet the needs of piscivorous birds such as the rare bittern (*Botaurus stellaris*) rather than protection of the fish themselves.

A further example of this disconnection is, perplexingly at least to my mind, amongst people who claim to be vegetarians but who eat fish – so-called pescatarians – as if fish were not in fact animals. I am sure that no deception is intended, but the conceptual disconnect is starkly evident.

The only time that the protection of fish species is taken seriously is when a few threatened species are scheduled for protection under nature conservation legislation. Protected species include Britain's three species of lamprey (in the family *Petromyzontidae*), two species of shad (family *Clupeidae*) and various members of the salmon family (family *Salmonidae*, including two whitefish species) or, in the case of the Atlantic salmon (*Salmo salar*), one of the UK government's former sustainable development indicators.

## 2.3 BELOW THE SURFACE

We probably all remember the school-day experiment of putting a stick or glass rod in a beaker of water and seeing it apparently bend. Light refracts as it passes through the surface of water, the differential densities of air and water acting exactly like a prism to change the pathway of photons. This has a range of interesting effects on fish biology, to which we will return later in this book. But it is also a good analogy with the ways in which public perception often halts, or at least changes, at the surface film.

As a scientist and angler, I am often astonished that people who walk their dogs or else pass by waters on a regular basis often fail to notice signs and sights of fishes and other river life that to me are blindingly obvious.

## 2.4 FISH AND FISHES

When are fish 'fish' and when are they 'fishes'? Linguistically, there is no firm distinction between the terms. However, they do tend to differ as a matter of established protocol.

'Fish' is most commonly used as the plural form of the word fish. For example, you might say that you saw fish in the river when looking over the village bridge, or that you bought fish from the fishmonger's shop. However, ichthyologists (scientists with an interest in fish) generally reserve the term 'fishes' for multiple different species, for example, noting that two freshwater fishes from the stickleback and tubesnout family (*Gasterosteidae*) occur in British fresh waters, with a further one a marine species. There is also the Biblical references to 'loaves and fishes', implying that the masses were fed a mix of species. By contrast, in a pond holding an abundance of just three-spined sticklebacks, we might accurately say that it was 'full of fish'.

## 2.5 GAME AND COARSE FISHES

One of the most frequent pair of terms that you may encounter segregating the fishes of Britain's fresh waters is a distinction between 'coarse' and 'game' fishes. These terms came into being during the late nineteenth century, prior writings referring simply to fish and the art of fishing. Prevalent understanding is that use of the word 'coarse' was related to the value of the fish as food items, as few British freshwater fishes other than members of the salmon family (that we will meet shortly) make good eating. In an era of less reliable food supplies, 'coarse' fish species such as wild-caught common bream and roach or pond-reared carp, tench and pike were

**A shoal of roach in an abandoned quarry.**

common fare albeit requiring significant preparation. Innovations in transport and preservation from the middle of the nineteenth century saw a progressive switch away from locally available coarse fishes in favour of tastier and now more widely available fresh sea fish and salmon, with aquaculture also substantially increasing the availability of more palatable species.

A schism was also happening in the world of angling, particularly the pace of development of fly fishing, that preferentially resulted in capture of 'game' fishes such as trout and salmon. Social changes too were progressing apace, as increasing wealth and access to travel enabled by railways meant that the emerging middle class could travel to cleaner, remote waters such as those found in Scotland to pursue salmon as a privileged field sport, along with hunting and shooting. This deepened the perception of 'game' species that could be hunted as distinct from 'coarse' fishes generally caught by more sedentary means.

As we will see, some species such as the grayling blur this apparently clear distinction, and today the fly angling sportsman may as often hunt 'coarse' fish species. In essence, the distinction in not useful, and not really distinct at all, but survives from former centuries as a legacy of diet, privilege and class division.

## 2.6 SHOALERS AND LONERS

Another interesting distinction that casual observers may discern is between fishes that are gregarious and those that are solitary for most of their lives. The terms 'shoal' and 'school' are generally used, the former often for aggregations of fish with a more distinctly structured formation with fish swimming more or less at equal distances apart and responding as a group, whilst a school is a far less formal arrangement. Again, this aggregation behaviour may not be clearly demarked. So why should fish shoal, or form schools?

One of the principal advantages of forming a tight shoal is that it reduces the potential for individuals to be picked off by predators. The shoal superficially resembles a bigger organism too large to interest the predator or else appearing as a threat. The darting behaviour within the shoal also potentially confuses the predator in its efforts to select any individual. A shoal also has more eyes and other senses at its disposal to pick out concentrations of food items or potential threats in the environment. Whilst some fish shoals have internal hierarchies amongst members, this does not seem to be the case for British freshwater species. Shoaling behaviour is also far more pronounced in younger life stages, smaller and more vulnerable fish fry benefitting from safety in numbers.

By contrast, fish that develop solitary behaviour as adults tend to be predatory, such as the pike which avoids others of its species to prevent competition for food, including risks of cannibalism. Other solitary fishes grow large including voracious omnivores such as common carp and wels catfish, which would compete for food if too frequently grouped closely together.

## 2.7 A CRAZY LITTLE THING CALLED FISH

So is a fish a 'thing' or an 'article'? Or is it something entirely different? A fascinating court case in 1954 revolved around exactly this legal nuance, in relation to the legality of the then-common practice of rag and bone men offering a goldfish as a reward for waste material picked up on their rounds. Was a goldfish, in legal terms, an 'article' or something else entirely? I quote here from a lovely and informative little book that has been with me from childhood: Kenneth Mansfield's *Small Fry and Bait Fishes: How to Catch Them* (1958):

*On January 19th, 1954, in the Queen's Bench Division, the Lord Chief Justice and two other justices ruled that a goldfish was not an article. They heard a case charging that the person before them, 'being a person who collected rags and was then engaged in collecting such articles, he delivered a certain article, namely a goldfish, to a boy under the age of 14 contrary to section 154 of the Public Health Act, 1936'.*

*In an attempt to determine whether a goldfish was an article or not, six definitions were read out from different dictionaries—and the Lord Chief Justice remarked that it seemed hard on a rag and bone man if he had to look up the dictionary to see what he could do and then found about ten definitions of the word.*

*It was pointed out that the rag and bone men who gave away goldfish knew the fine distinction between a thing and an article and operated accordingly. They presented goldfish, which were things, but insisted that the boys who received them brought their own containers, for if they also gave away a bowl or a jam jar they would be infringing the law by giving away an article.*

*Supporting the rag and bone men's argument, the Court ruled that a goldfish was a thing but not an article.*

So, according to my reckoning, this mean that the book you are reading is an article, but it is written about things (fish). That is that weighty matter settled then!

# WHAT HAVE FRESHWATER FISHES EVER DONE FOR US?

## 3

Before delving into the inner worlds of fishes, and reflecting on the fact that fish are so often underappreciated, it is instructive to consider the many ways in which our heritage of freshwater fishes benefits us.

A bit like the famously lampooned question, 'What have the Romans ever done for us?', we are perhaps over-familiar with the significant benefits that Britain's freshwater fishes, as indeed the natural world as a whole, bestow upon us. So this chapter is about the diversity of services by which freshwater fishes support human wellbeing. Fishes, in fact, do rather a lot of things for us, many of which we have perhaps failed to appreciate in these terms and then to factor into the ways in which we interact with, protect and manage the world.

## 3.1 THE MANY HUMAN USES OF FRESHWATER FISHES

Britain's freshwater fishes are, or have been, directly exploited as physical resources by people in a wide diversity of ways, from food to medicines and ornamental and other resources.

### 3.1.1 Fish as food

Data on the size of the global market for 'seafood' from both fresh waters and oceans is elusive. A 2010 report by Global Industry Analysts, Inc., projected that the global market for seafood (including frozen, canned and fresh fish from both capture fisheries and aquaculture) would reach US$371.3 billion by 2015. Subsequent reports by various bodies have noted continuing strong growth at least up to 2030, with aquaculture overtaking wild capture as the predominant source. Of this global contribution, a 2016 report* observed that catches from fresh waters are often overlooked, yet hundreds of millions of people around the world benefit

A huge diversity of fishes and other aquatic life is found in fish markets around the world. (Image © Dr Mark Everard.)

* McIntyre, P., Reidy Liermanna, C.A. and Revenga, C. (2016). Linking freshwater fishery management to global food security and biodiversity conservation. *Proceedings of the National Academy of Sciences of the United States of America*, 113(45), pp. 1,2880–1,2885. doi: 10.1073/pnas.1521540113.

from low-cost protein, recreation and commerce provided by freshwater fisheries, particularly in regions where alternative sources of nutrition and employment are scarce. The numbers are substantial, the report finding that freshwater fisheries provide the equivalent of all dietary animal protein for 158 million people, with poor and undernourished populations particularly reliant on inland fisheries compared with marine or aquaculture sources.

In contemporary Britain, the fish we eat are generally marine species, or else farmed trout, salmon and other species, that may have been transported long distances under refrigeration or in a frozen state. Yet before the advent from the middle of the nineteenth century of refrigeration and transport technologies, rail in particular, the process of conveying fresh fish inland from the coast was onerous, time-consuming and costly. In former times then, locally wild-caught or bred fishes, both game and coarse alike, were a more common staple food source at any distance from the shore. Angling in that era was as much about subsistence as enjoyment. Recipes for most freshwater fish species can be found in Izaak Walton's 1653 book *The Compleat Angler* and various other classic texts. This remained true right up to the inter-war period of the twentieth century, including, for example, Hugh Tempest Sheringham's marvellous and evocative writings from either side of the First World War. The angler's quarry generally ended up being carried home in a creel (a traditional angler's basket), and the old angling term for a successful capture, to 'kill of a fish', was no mere metaphor. My own books *The Complete Book of the Roach, The Little Book of Little Fishes* and *Dace: The Prince of the Stream* contain various references to, and reproductions of, published recipes for everything from roach to dace, and not overlooking smaller fishes such as gudgeon, minnows and bleak. Just one example of a classic use of a common freshwater species, as recorded in *The Compleat Angler*, is the Minnow Tansy:

*And in the spring they make of them excellent Minnow-tansies; for being washed well in salt, and their heads and tails cut off, and their guts taken out … being fried with yolk of eggs, the flowers of cowslips and of primroses, and a little tansy; thus used they make a dainty dish of meat.*

Of other British freshwater fishes, Walton has much to say, including of the grayling:

*…and Gesner says, that in his country, which is Switzerland, he is accounted the choicest of all fish. And in Italy, he is, in the month of May, so highly valued, that he is sold there at a much higher rate than any other fish.*

Also that:

*The Perch is of great esteem in Italy, saith Aldrovandus: and especially the least are there esteemed a dainty dish. And Gesner prefers the Perch and Pike above the Trout, or any fresh-water fish: he says the Germans have this proverb, 'More wholesome than a Perch of Rhine': and he says the River-Perch is so wholesome, that physicians allow him to be eaten by wounded men, or by men in fevers, or by women in child-bed.*

And of bream (no distinction was made in *The Compleat Angler* between the two British species: common bream and silver bream):

*But though some do not, yet the French esteem this fish highly; and to that end have this proverb "He that hath Breams in his pond, is able to bid his friend welcome"; and it is noted, that the best part of a Bream is his belly and head.*

Commercial fisheries for some wild British freshwater fish species can be substantial. For example, licensed netsmen taking sea trout primarily from northeast England generated sales estimated at £200,000 in 2003.

Nevertheless, some freshwater fishes remain familiar fare on our tables today, though more often from farmed rather than wild-caught stocks. Increasingly commonly imported freshwater fishes include Pangasius catfish (mainly comprising two species: the 'tra', *Pangasianodon hypophthalmus*, and the 'basa', *Pangasius bocourti*), 90% of which is farmed in Vietnam and sold to global markets as white fillets under such western-friendly names such as 'panga', 'basa' and 'river cobbler'. Salmon and trout make up the bulk of freshwater fish species consumed in Britain, though there is a thriving market in coarse fish species to serve eastern European consumers. In not-so-recent history, many species of freshwater fish were taken as food prior to the Second World War when local food was a far more important constituent of the national diet. Perch and pike were in far more recent times commonly taken for the table, as some still are. A relatively recent introduction to the British freshwater fish fauna, the zander, is widely consumed across its northern Eurasian distribution and has a firm, tasty flesh.

An interesting project initiated in response to food shortages during the Second World War was the commercial harvesting of perch from Lake Windermere, in the English Lake District in Cumbria. As food supplies came under pressure due to blockades, the Freshwater Biological Association (FBA), headquartered on the banks of England's largest lake, sought to make a contribution to the war effort by exploring the harvesting of stocks of fish in freshwater lakes. Perch were amongst the most abundant edible freshwater fish in British lakes. After experimentation to determine the migration patterns of perch and pike in Windermere, the FBA discovered perch could be caught in numbers using appropriately positioned fish traps. An original array of three hundred commercial-scale perch traps was constructed in the spring of 1941 from fencing wire covered with mesh wire netting. These nets were dipped in black tar-varnish to increase both their catches and their life, and covered in with a funnel entrance and a

The perch, a handsome freshwater fish that was the focus of commercial harvesting in the English Lake District as part of the Second World War food security effort.

small door to remove the catch. To supply potential markets, a canning factory in Leeds agreed to buy the perch and process them into sardine tins to be sold as '*Perchines – Lakeland Perch, Britain's most lovely and tasty fresh water fish*'. Reports suggest that perch were far from in their tastiest state by the time they had spawned, been trapped and transported to Leeds and then canned, though sales were promoted by packing them in Yorkshire relish and tomato ketchup. In reality, the strongest market incentive was that wartime food rationing coupons were not required to purchase these home-produced tinned fish. The initial relative success of the Windermere Perch fishery led to the Ministry of Agriculture and Fisheries organising the trapping of perch on several other Cumbrian lakes as well as a handful of other lakes and reservoirs in England and Wales, and also some Scottish lochs and reservoirs. None of these ventures lasted more than a year or two, with renewed access to more palatable fish at the conclusion of the war serving as a 'coup de gras' for this nascent British commercial perch fishery industry. More than seventy years later, harvesting activities continue at Windermere, but they have since been solely as part of a long-running scientific study into fish population structure and dynamics.

Locally across Britain, fish produced for the table from some freshwater fisheries have had regional and sometimes far wider significance. These include more immediately evident cases such as the considerable market premium for wild-caught Atlantic salmon and sea trout. Another now-largely-defunct example is the former Windermere Charr fishery, now more significant culturally than gastronomically. Potted charr from the English Lake District in Cumbria was long considered a great delicacy, cottage industries between the seventeenth and nineteenth centuries exporting this product the length and breadth of the country, often in specially designed pots or dishes. Initially, this deep-water fish was caught by netting.

However, a novel technique of rod-and-line fishing using long poles attached to the sides of boats on Windermere appears to have been first developed around 1840, Francis Day in his 1887 book *British and Irish Salmonidae* reporting that '*It will be between forty and fifty years ago since a Mr Spencer, from Manchester, first introduced the plumb line into the Lake District... his success, season after season, speedily induced imitators, and the plumb line did not take long to become established*'. Arctic charr and this unique method of fishing for them are synonymous with the Lake District. The coda to this point though is that plumb line fishing continues today at a very small scale, though now mainly as a tradition and curiosity rather than as a commercial venture beyond a niche angling market.

Other British freshwater fish species that have become local delicacies or the subject of regional traditions include European eels, once abundant though now increasingly scarce. Jellied eels are one such traditional British dish, reputedly originated in the eighteenth century primarily in the East End of London. Jellied eels are produced by boiling chopped eels in a spiced stock that is then allowed to cool and set, forming a jelly, and eaten cold. 'Eel pie, mash and liquor' is another food that was once a traditional London working-class food dating back centuries, with many 'mash and eel' shops established in the capital since the nineteenth century and many still found in East and South East London and some parts of Kent and Essex. Whilst the eel and mashed potato elements are no mystery, the exact constitution of the green 'liquor', basically a savoury parsley sauce, is a secret guarded closely by chefs. Smoked eels are certainly popular in continental Europe, if not quite so much in Britain, as are baked eel sections.

As we will discuss in greater detail in the chapter 'The curious world of Britain's freshwater fishes', eels, and glass eels in particular, are a very highly prized delicacy in Asia. Indeed, they are so prized that the glass eel stage of the European eel life cycle commands hundreds of pounds Sterling per kilogramme, prices rising with the estimated eel population decline of 90%–95% over the last forty-five years that has resulted in the classification of the species today as 'Critically Endangered'.

Other freshwater fish species found in Britain used to produce gastronomic delicacies are the vendace, used to make a form of caviar across Scandinavia where the species is common, and the burbot, used for the same product in Finland. The vendace is, however, scarce, found in generally inaccessible depths in fragmented scattered still waters and also legally protected in British waters, whilst the burbot is extinct here. So these two species clearly have no value for this purpose in our home fresh waters! Of the common carp, Walton says, '*But it is not to be doubted but that in Italy they make great profit of the spawn of Carps, by selling it to the Jews, who make it into red caviare; the Jews not being by their law admitted to eat of caviare made of the Sturgeon, that being a fish that wants scales, and, as may appear in Leviticus xi., by them reputed to be unclean'*. This is corroborated by Charles Stewart in his 1817 book *Elements of the Natural History of the Animal Kingdom*, which notes of the common carp that '...*of its spawn is made a species of caviar...*'

Freshwater fishes have also been farmed for millennia, with aquaculture techniques constantly advancing. Lucius Licinius Murena, consul in the late Roman Republic in 151 BCE, is said by Pliny to have invented fishponds. In fact, his cognomen (an extra personal name given to an ancient Roman citizen serving more as a nickname than in any formal sense) is derived from the Latin word *murenae*, meaning a lamprey or an eel. (Perhaps by coincidence, 'Lucius' is the name of the genus of the pike.) The identity of these fishes as eels makes a lot more sense than lampreys in the context of aquaculture, a point to which we will return. Chinese aquaculture also has a very long history, much of it now lost in the mists of time.

Two species of freshwater fishes lay a claim as the most globally widely produced species in aquaculture: the silver carp and the grass carp. The silver carp (*Hypophthalmichthys molitrix*), a variety of Asian carp native to north and northeast Asia, has been cultivated in China for centuries and is now widely farmed and introduced into the wild globally, though none occur in British waters. The grass carp too has been farmed in the Far East for centuries and is also widely cultured and introduced globally, including occurring in British fresh waters, though as a non-breeding sporting species rather than as food. Of the many other freshwater fish species produced in aquaculture, rainbow trout and brown trout dominate the market in Europe. Trout farming was introduced to the UK in the 1950s by a Danish entrepreneur. Almost 360 trout farms are found across Great Britain at the time of writing. The rainbow trout, native to rivers of America's Pacific coast, is generally hardier and faster-growing than our native brown trout. Rainbow trout are consequently more common in British aquaculture, where production is around 16,000 tonnes annually. Brown trout are also produced in significant quantities, with smaller volumes of blue trout (a cultivated colour variant of rainbow trout) and fewer golden trout (*Oncorhynchus mykiss aguabonita*: a sub-species of rainbow trout native to California). The market for farmed Atlantic salmon is also strong and growing, though these fish are usually cultured in sea cages.

Aquaculture in British fresh waters is far from recent. Aquaculture practices spread to great estates and monasteries, with stew ponds becoming increasingly common as a source of fresh year-round meat during the Dark Ages (the period following decline of the Roman Empire, roughly from the tenth and eleventh centuries). Romans were known to keep eels in fish ponds (vivaria), ready to hand for the table when required (discussed in more detail in Chapter 10). In Europe, stew ponds were regularly recorded in the earliest records of predominantly Roman Catholic religious orders, particularly after Papal decrees classifying fish as acceptable 'Lent meat'. The importance of fish in monastic life throughout the Middle Ages is evident through many surviving documents. At this time, the farming of fish in finger ponds attached to rivers and estuaries became prevalent, though these were prone to flooding, as well as in more isolated stew ponds. The eleventh century *Domesday Book* recorded large numbers of private *vivariae piscinae* on estates across England, attracting a tax comprising part of the crop which, at the time, comprised mainly pike and eels, as well as bream. Peasants were at this time landless, deprived of ownership of stew ponds and access to fish from other waters owned by ruling elites. Few waters were deemed common property, and poaching was not infrequently met with the death penalty. Fresh fish in this time therefore became a rare and privileged commodity. Fish retained in stews year-round might have been observed breeding in the spring. It would have been a short innovation from that awareness to the digging of separate ponds for storing fish over winter, breeding them in the spring and fattening them in the summer and autumn.

The common carp is mentioned by name by King Charlemagne in his book *Capitulare de Villis*, written in the year 812, indicating that simple fish husbandry practices arrived with the species from Asia into Europe in the ninth century. The spread of many species beyond their native ranges, both internationally and within Britain and other regions, has often been driven by their exploitation in aquaculture. Common carp were reputed to have been introduced to Britain in the late fifteenth or early sixteenth centuries by monks as a source of food, Izaak Walton reciting Sir Richard Baker's verse that '*Hops and turkies, carps and beer, Came into England all in a year*'. However, carp may have been recorded in stew ponds in England from the fourteenth century. Monks are generally considered to be Britain's first fish farmers, benefitting from levels of education and awareness of innovations overseas that were beyond the reach of the mass of the population. Monks were required to observe a diet entailing eating fish throughout all of Lent and on other fast days, which clearly created problems for those not living near the sea. Monasteries also had adjacent land, and on many of them, 'stew ponds' were developed to produce fish. Famously, and as reported by Baker and Walton, carp were introduced for this purpose as they are tolerant of poor water quality and variable environmental conditions, will eat almost anything and also grow large and rapidly. In many ways, common carp are like pigs – in my book *Fantastic Fishes* I refer to them as '*pigs with fins*' – explaining why they are so widely introduced globally as a food source. (Pork is the most widely consumed meat in the world, followed by poultry, beef and mutton.) Common carp have the added benefit that they do not smell like their terrestrial comparators, enabling the monks to stay clean for prayers. The basics of carp production were not complex. Ponds were dug with a ramped edge, the first of them filled with water and stocked with carp, with a second pond allowed to be grazed by livestock that manured the bed. When alternate ponds were drained down annually, enough fish were taken to feed the monks with breeding stock transferred to the second pond, whilst the first was again manured, meaning that plants and invertebrates produced within the pond averted any need for intensive feeding of the fish.

THE COMPLEX LIVES OF BRITISH FRESHWATER FISHES

The common carp, widely spread from its native range for aquaculture, was introduced into Britain at some time during the fourteenth to the sixteenth centuries.

Fish taken for eating were transferred to a clean third pond, feeding only on insects landing on the pool's surface, to retain the fish until they were required and to purge their flesh of its muddy taste. Some remnant ponds, mainly surviving today as indistinct depressions near old monastic sites, are now largely all that remains following the dissolution of monasteries in the English Reformation (a series of events in sixteenth century England by which the Church of England broke away from the authority of the Pope and the Roman Catholic Church).

As an addendum to this tale, there is a view that tench rather than common carp were the first fish farmed in this way by monks. Whilst the tench is a native species, the natural range of the common carp is from the Danube River extending to the Black Sea, Caspian Sea and Aral Sea and, though a very hardy fish, would have been transported over long distances. Evidence supporting this claim is scarce and contested. Tench were not, from literature of the time, generally considered a good fish for eating, perhaps as much related to their tiny scales and superficially scaleless appearance conflicting with religious diktats, though the tench did not formerly occur naturally in the geographic range where these tenets were written. However, tench did used to have medicinal uses, as we will see in Section 3.1.3. Common carp certainly subsequently became a species of choice, probably for their large size and fast growth rate.

The roles that Britain's freshwater fishes have played in food provision, and their supporting traditions and economies, have been and remain significant. However, as discussed previously in this section, the long historic exploitation of fish as food and for recreation appears to have had a strong influence on their perception as something apart from other 'wildlife'. In Britain, as elsewhere in the world, this group of animals is seriously overlooked as an integral element of our wildlife heritage and in conservation management.

The three-spined stickleback, one of two British freshwater stickleback species that has formerly been harvested for fertilising fields.

### 3.1.2 Fish as fertiliser

Captain T. Williamson's 1808 book *The Complete Angler's Vade-Mecum* records that sticklebacks were *'sold by the bushel, as manure, both in Lincolnshire and in Cambridgeshire'*. This observation is corroborated by Thomas Frederick Salter's 1815 book *The Angler's Guide*, which reports that:

> *Pricklebacks are frequently used, in Lincolnshire, for manure, being always very numerous in the fens; but sometimes, they become so numerous as to make it necessary to separate and find new situations, which happens once in eight years, upon an average; during which migration, part of the river Welland is almost choked with them, at which time they are collected in nets, sieves, baskets, &c., to the amount of cart loads, and spread on the land as manure, and, I am informed, fertilize it extremely.*

This is of course unfortunate for the sticklebacks concerned. However, then present in such clear abundance, the harvesting of stickleback serves as an efficient and natural means to transfer nutrients from watercourses into adjacent croplands.

### 3.1.3 Medicinal uses of Britain's freshwater fishes

For centuries, the tench has been known as the 'doctor fish'. In *The Compleat Angler*, Izaak Walton's Piscator (the angling expert) writes of *'The Tench, the physician of fishes'*. He continues, *'In every Tench's head there are two little stones which foreign physicians make great use of... Rondeletius says, that at his being at Rome, he saw a great cure done by applying a Tench to the feet of a very sick man. This, he says, was done after an unusual manner, by certain Jews'*. To this, he adds that, *'Well, this fish, besides his eating, is very useful, both dead and alive, for the good of mankind'*. Early physicians prescribed the touch of

The tench, or 'doctor fish', has been ascribed a range of medicinal properties.

a tench as a cure for the ague (malaria or other illnesses involving fever and shivering), the fish applied like a large plaster to the soles of the feet of the afflicted patient to draw out fever.

It is not just the benefits to human health for which tench were renowned. Walton's Piscator again tells us *'that the Tench is the physician of fishes, for the Pike especially, and that the Pike, being either sick or hurt, is cured by the touch of the Tench. And it is observed that the tyrant Pike will not be a wolf to his physician, but forbears to devour him though he be never so hungry'.* Walton attributes this to the fish carrying *'a natural balsam in him to cure both himself and others, loves yet to feed in very foul water, and amongst weeds'.*

Neither modern science nor observations on the diet of Pike bear any truth in these established myths, though the legend of 'doctor tench' persists, including in its alternative common name of the 'doctor fish'. Tench are, however, far from unique as freshwater fishes to which therapeutic benefits are ascribed.

Walton also reports that perch *'...have in their brain a stone, which is, in foreign parts, sold by apothecaries, being there noted to be very medicinable against the stone in the reins'* (an archaic word for kidneys). Of the common carp, Walton also says that *'The physicians make the galls and stones in the heads of Carps to be very medicinable'.*

Various species of fish, as indeed many other animals and their parts, are used in traditional medicinal practices throughout the world. For example, medicinal fish markets are commonly found in Thailand's major cities. Fish and shrimps jostle with better known plants, mammals, reptiles and other groups of organisms in Chinese traditional medicine. Many traditions of oriental medicine substantially predate our own, show considerable sophistication and have commonalities amongst cultures. However, exploitation of fish in medicine is far from restricted to the Far East.

In the fenlands of East Anglia, arguably Britain's own Far East, many local remedies were invoked to ward off indigenous diseases. Dried eel skin is one such supposed therapeutic agent formerly commonplace in Suffolk, worn in the form of garters around the thighs to ward off rheumatism and cramp, which were prevalent in the wet climate of drained marshes. Scottish medical tradition also made use of eel skins, which were wrapped around the fingers to ward off cramp. Eel skin worn on the naked leg was also believed in the North Country to be sure to prevent cramp when swimming. Another old English belief was that to cure a man of drunkenness, a live eel should be placed in his drink. There may be wisdoms in such practices that are often dismissed as superstitions as, thousands of miles away, the related Indian mottled eel (*Anguilla bengalensis bengalensis*) is ascribed similar qualities, and mucous from live Indian mottled eels is also mixed with rice or wheat flour as a medicine for arthritis.

Potions of all kinds can, of course, be both medicines and poisons. So it is perhaps not surprising that Britain's freshwater fishes can contain poisons. In Japan, we have the famous example of the greatest gastronomic delicacy, fugu. Fugu are pufferfish, the 121 species of pufferfish being the second most poisonous vertebrates in the world after the golden poison frog. Pufferfish produce the potent neurotoxin tetrodotoxin (TTX), which is not broken down by cooking or in waste parts discarded from preparation and for which there is no known antidote. Preparation of fugu in restaurants requires many years of apprenticeship to train a master chef so that diners can enjoy the fish without risk of death, a risk so high that fugu is banned throughout the European Union. That said, TTX may have a beneficial use as an analgesic agent to help patents deal with chronic cancer pain where no other pain medication has worked. Britain's fresh waters lack anything quite as potent as Japan's estuarine and coastal pufferfishes. However, the roe of the barbel is a well-known example of a poisonous

The European eel, another elusive British freshwater fish ascribed a wide range of medicinal properties.

fish part, though the flesh of the barbel is said to be good eating if a bit bony. The roe, by contrast, causes vomiting and diarrhoea, of which Frank Buckland states in his 1880 book *Natural History of British Fishes* that '*The barbel is not good food, nevertheless the Jews eat him during their holidays. The eggs of the barbel are said to be poisonous, and to produce the same symptoms as belladonna. Fishermen about Windsor have a horror of barbel's roe, so there must be something in the story*'. In the classic 1653 book *The Compleat Angler*, Izaak Walton's Piscator tells his student, '*But the barbel, though he be of a fine shape, and looks big, yet he is not accounted the best fish to eat, neither for his wholesomeness nor his taste: but the male is reputed much better than the female, whose spawn is very hurtful, as I will presently declare to you*'. Later in the book, Piscator adds that '*the spawn of a barbel, if it be not poison, as he says, yet that it is dangerous meat, and especially in the month of May; which is so certain, that Gesner and Gasius declare it had an ill effect upon them, even to the endangering of their lives*'.

### 3.1.4 Decorative and other uses

Another British freshwater fish that has been put to human use, beyond its angling and gastronomic values, is the bleak. Until as late as 1870, the rivers Trent and Thames in England played host to substantial commercial fisheries capturing many millions of bleak. They were caught for their scales, from which the silvery guanine crystal deposit was extracted for making pearl essence *(essence d'orient)* and artificial pearls. To produce these valuable products, bleak scales were scraped from the flanks of captured fish. The stripped bleak were subsequently discarded, or else sold on as a cheap food. After the mud and slime has been cleaned from the bleak scales, they were soaked in water to loosen the pigment, which was dislodged and allowed to accumulate at the bottom of the vessel. Reports indicate that between 4,000 and

The bleak, formerly and in Macedonia currently, used for the manufacture of 'bleak pearls'.

5,000 bleak were required to produce 100 grammes of *essence d'orient*, which was therefore an expensive commodity.

Bleak pearls were initially produced by mixing the *essence d'orient* with wax or dense fat. However, subsequent innovations in France enabled the production of far more durable bleak pearls by drawing the *essence* up into thin glass tubes, which were then blown into hollow glass beads of varying forms and sizes. Roach, dace and some other silver fishes were also used for scale production. However, bleak were abundant and provided the best quality *essence d'orient*, and they were therefore very much in demand.

Bleak pearls were traditionally used in the characteristic garb of London's cockney 'pearly kings and queens'. However, they had far wider applications in the bead trade, for example, in necklaces and earrings. At the height of their fashionable appeal, ornaments manufactured with bleak pearls bore the name 'patent pearl'. The demand for this product was so great that the price of a quart (2 pints or 1.13 litres) measure of fish scales varied from one to five guineas (£1:05 to £5:25: assuming 1870 prices, this is more or less equivalent to £115 to £575 today). This practice is said to continue in continental Europe, particularly in eastern Europe, with by-products used for fertiliser and animal food, although evidence supporting this is elusive. However, the bleak pearl industry was subsequently largely replaced by 'Roman pearls', the nacre for which was instead derived from the swim bladders of *Atherina* (types of sand smelt) largely caught from the Mediterranean. Although British estuaries host one species of *Atherina*, the sand smelt, no such industry took off here, with foreign production, including plastics, now dominating the market.

Other species of British freshwater fishes have been put to a range of practical uses. Of the common carp, Charles Stewart records in his 1817 book *Elements of the Natural History of the Animal Kingdom* that '...its bile is used as a green pigment; of its spawn is made a species of caviar, and of its swimming-bladder isinglass'. Other of the many uses of freshwater fish parts include the historic use of oiled eel skin as an opaque substitute for window glass.

## 3.2 THE ROLES BRITAIN'S FISHES PLAY IN FRESHWATER ECOSYSTEMS

Britain's freshwater fishes clearly have many uses and, as we will subsequently see, they also contribute significantly to cultural enrichment. However, we should not overlook their many roles in keeping the ecosystems, of which they are integral elements, functioning and producing other less direct human benefits.

### 3.2.1 Freshwater fishes as integral links in ecosystem functioning

Freshwater fishes, for example, serve as vital links in food webs, cycling nutrients, other chemicals and energy that maintain not only freshwater ecosystems but the terrestrial ecosystems that interact with them. As grazers, predators or prey, omnivores and herbivores as well as consumers of detritus, our diverse freshwater fish species and their varying life stages maintain and promote the cycles of nature, the vitality of aquatic systems, links with terrestrial systems and the many human benefits that accrue from them. Take away the fishes, and these functions suffer.

### 3.2.2 The roles fishes play in life cycles of other organisms

Fishes also play many other roles in the life cycles of other species. As one example, they are a food source for many other organisms, at the same time also playing key roles as predators, maintaining population sizes and health.

An extreme example of the role of freshwater fishes in the life cycles of other species will be touched upon when considering their sex lives, and in particular the roles that brown trout and Atlantic salmon play as hosts of the parasitic *glochidia* larval stage of the pearl mussel (*Margaritifera margaritifera*). Female pearl mussels produce eggs in their mantle cavities, fertilised by sperm released into the water by adjacent male molluscs. A single female mussel may release as many as four million glochidia larvae, which exit via the siphon tube after hatching. A fortunate few larvae are inhaled by chance by a trout or salmon, the hinged larva's body snapping shut on the gills of the fish and forming a cyst that continues to grow and metamorphose. The fish provides the glochidia with nutrition, dispersal and protection until the following spring or early summer. Then, the seed mussel drops off to commence an independent life in clean, fine river gravel, living as long as 120 years. The pearl mussel is now critically endangered because of multiple pressures, significantly including decreasing water quality and sedimentation, as well as their close dependence upon strong populations of salmonid fishes to complete their remarkable life cycle.

The pearl mussel, an exceedingly rare and vulnerable bivalve mollusc of British fresh waters with fascinating interdependencies with freshwater fishes essential for it to complete its life cycle.

### 3.2.3 The roles fishes play in parasitic life cycles

Many fish species also play crucial roles in the life cycles of parasites. It is often not in the interests of parasites to cause undue harm to its host, the death of the host also shortening the parasite's life. The exception to this is parasites that have a vested interest in their host being consumed to complete the parasitic life cycle. In this latter case, the fish is known as an 'intermediate host', eaten by a 'final host' (most commonly a warm-blooded bird or fish) where it can then metamorphose into its adult, sexually mature form. There are many examples of this type of complex, multi-host life cycle.

One such example is *Diplostomum*, a genus of digenean flukes (trematode worms usually equipped with two suckers), many species of which afflict the fishes of the world. The sexually mature parasite is found in the gut of various piscivorous bird species, including herons, gulls and terns. Eggs released by the mature parasite exit the host bird as it defaecates, landing in water and hatching into microscopic *miracidium* larvae that seek out and penetrate the skin of lymnaeid snails (such as the great pond snail *Lymnaea stagnalis*). Inside a suitable host snail, the *miracidia* (plural of *miracidium)* metamorphose into *cercaria* larvae which reproduce in vast numbers in the hepatopancreas (a large organ combining the functions of both liver and pancreas). *Cercariae* subsequently metamorphose into *metacercariae*, which break out of the snail into the surrounding water, seeking out and actively penetrating the skin of a suitable fish intermediate host. Once they have burrowed into the fish, the *metacercariae* migrate along the blood vessels to the eyes, metamorphosing again into hard, resistant cysts known as *diplostomula* which accumulate in the lens. Heavy infestations can totally blind the host fish, lower levels of infection also reducing the fish's capacity to detect and evade aerial predators. The intermediate fish is thereby rendered more susceptible to being caught by a final bird host, in the gut of which the parasite embeds itself to complete its life cycle.

A three-spined stickleback bloated by an intermediate parasite, progressively disabling the fish such that it is increasingly vulnerable to predatory birds that are the parasite's final host.

Other parasites using fishes as intermediate hosts may incapacitate their hosts in other ways. For example, the tapeworm *Ligula* grows in and distends the body cavity of a host fish, progressively debilitating and disrupting the camouflage of the fish and increasing its susceptibility to predation by a warm-blooded final host.

## 3.3 CULTURAL VALUES OF FRESHWATER FISHES

In addition to their utilitarian values and contributions to the functioning of ecosystems supporting a diversity of other human needs, many of Britain's freshwater fishes play significant roles in our cultural wellbeing.

### 3.3.1 Angling and its many values

Angling, or fishing with an angle (or hook), is a sport enjoyed by millions in Britain. It is a sport with so many angles (pardon the pun), ranging from competition and economic activity to quiet relaxation, social inclusion and conservation management. This is not a book about sport fishing, but exploration of some of these dimensions and associated values is certainly instructive about the cultural importance of Britain's freshwater fishes.

A range of studies have quantified the economic importance of recreational angling in England and Wales as approximately equal to the contribution of farming to GDP (gross domestic product). That statistic alone indicates the scale of angling's significance. The 2019 report *Angling for Good: National Angling Strategy 2019–2024* stated that angling was '…*estimated to involve around 900,000 fishing in freshwater in England and Wales…*' with '…*significant economic impact… … freshwater angling in England in 2015 contributed £1.46 billion to the economy and supported 27,000 full-time equivalent jobs*'. Also reported in this survey of anglers, '*72% of respondents said that it helped to keep them healthy, 62% saying angling was one of their ways of being physically active and 25% saying it was their main way of being active*'. Also, '*70% said that angling helped them de-stress. 58% of respondents would access nature less often or not at all if they did not go angling, 70% would visit rural areas less often*' and '*57% of those surveyed had been involved in environmental improvement volunteering*'. A 2007 Environment Agency report, 'Economic evaluation of inland fisheries', concluded that '*Freshwater Angler gross expenditure across the whole of England and Wales was £1.18 billion. Household income of £980 million and 37,386 jobs were generated across England and Wales*'. It is generally concluded that angling was worth at least £2.5 billion to England and Wales and more than £3 billion to the UK as a whole, taking account of fresh and saline waters. A figure of £4 billion is more often quoted today. However, these figures may well be an underestimate when the value of angling in Scotland is considered, accounting for a whole set of linked trades such as ghillying and river keeping, catering, hospitality and tourism. A 2007 survey found that angling on the River Tweed was worth more than £18 million to the Border region's economy, supporting around 500 full-time equivalent jobs, whilst a Scotland-scale survey assigned a total value of £113 million, to which salmon contributed £78 million.

**Recreational angling has substantial economic and other values. (Image © Vince Cater.)**

It is important here to dissociate the value of live, wild fish as distinct from the market price of a dead fish. For example, a commercially caught wild British salmon of 3 kg might sell for £45, at a price to a commercial fisherman of £15 per kilogramme, with that fish clearly lost to the spawning stock in order to realise its value. However, if the same fish is caught by rod and line and returned alive (as most now are), then it can not only survive to reproduce, but would also contribute between £500 and £2,000 to local economies depending on where it was caught. Further value arises from additional reinvestment in river management by angling interests. As one example, a survey in 2005 on Hampshire's Rivers Test and Itchen found that anglers spent £3.25 million to fish that year, of which £3 million was reinvested into management of those rivers by fishery owners, supporting 120 full- and part-time jobs in the process. Two further examples published by the Environment Agency in 2010 revealed a wide range of economic values from sea trout habitat restoration on the River Glaven in North Norfolk, the second study assessing the installation of a 'buffer zone' to protect a formerly severely cattle-poached field margin on the upper Bristol Avon in North Wiltshire. Both of these 2010 studies ostensibly focused on fish and angling interests but, in both cases, the actual fishery benefits were dwarfed by wider water quality, erosion regulation, wildlife conservation, tourism, amenity, social relations and related benefits.

Further social benefits accrue from angling activities. A 2010 report 'Angling: A Social Research' identified significant contributions from recreational angling to economic impact and value, including roles in participation and social inclusion, health and wellbeing, tourism and the viability of rural communities in the UK. Teasing out robust values is elusive as much of angling is a 'hidden' private pursuit, unlike organised sports such as soccer or cricket. It also lacks a high media profile and has a relatively shallow associated body of published research. However, angling is generally perceived to be the largest participant sport in the UK, an Environment Agency report 'Public Attitudes to Angling 2005' calculating that 8% or 3.5 million people had been fishing in the fresh waters of England and Wales in the preceding two years. But significantly, it is also a pastime with involvement of or access by people across age, class, ethnic, income, physical ability or disability, gender and other divisions, offering a high degree of social inclusion and the diversion of the young from other potentially problematic activities and behaviours. Challenges remain as the majority of anglers today are older white males, but angling is recognised by government as having a significant potential role in enhancing 'community cohesion'. The 2004 Department of Health White Paper, 'Choosing Health', outlined overarching health priorities including tackling health inequalities, combating obesity particularly amongst young people and men, stopping smoking, reducing alcohol and substance misuse, and reducing teenage pregnancies, calling for active promotion of 'mental wellbeing', particularly in areas of high levels of deprivation. These are goals towards which the diverse potential impacts of angling and other outdoor participatory activities can contribute, including now well-established health effects arising simply from exposure to natural environments. Whilst these qualitative aspects are relatively poorly researched, many are reported as the most significant reasons why people go fishing, so their potential roles may be cumulatively highly significant for national wellbeing.

A further benefit from angling arises from the mass of interested observers the length and breadth of the country providing detailed and locally attuned, voluntary surveillance of Britain's fresh waters. As the Environment Agency's 2003 National Trout and Grayling Fisheries Strategy puts it, 'Indeed, anglers are often the first to raise concerns about issues such as pollution, over-abstraction or physical damage to the aquatic environment'.

### 3.3.2 Pet fish

Many people choose to share living spaces with living freshwater fishes, be that in tanks or ponds, or simply by spending time with them in their natural environment as fish watchers or anglers. A tank of fish in a dentist's waiting room has a calming effect, just as a pool of fishes may be central to Buddhist or other meditative spaces.

Pet fish are in fact the subject of a massive global trade. The 2007–2008 National Pet Owners Survey (NPOS), undertaken by the American Pet Products Manufacturers Association (APPMA), revealed pet ownership in the US at its highest recorded level with 71.1 million households in the US owning at least one pet (63% of the 113,707 million US households). These numbers include an estimated 142 million freshwater fish, substantially more than either cats or dogs, as well as 9.6 million saltwater fish. In Britain, the Pet Data Report 2017, undertaken for the Pet Food Manufacturers' Association (PFMA), revealed that 8% and 4% of UK households owned 'indoor fish' and 'outdoor fish', respectively, cumulatively accounting for 33 million fish, supporting a pet food market of £69 million. It seems that we have a subconscious need to find room for fish and the environments that support them in our daily lives.

Goldfish artificially bred in a wide variety of body and fin shapes, as well as colours, are commonplace in aquaria and garden ponds, as are golden orfe (a cultivated golden form of the orfe). So too are the highly prized koi carp, an often ornately coloured strain of the common carp. Koi can potentially command astronomical prices, particularly in Japan, but with a global market, the most beautiful individual specimens selling for up to US$2.2 million. Common carp have been bred in ponds for food for at least 2,400 years in China and spread westwards into Europe during the time of the Roman Empire. Christopher Currie's 1991 study *The Early History of the Carp and its Economic Significance in England* quoted from Cassiodorus (AD 490–585) that carp graced King Theodorus' court in Italy where *'From the Danube come Carp and from the Rhine Herring. To provide a variety of flavours, it is necessary to have many fish from many countries. A king's reign should be such as to indicate that he possesses everything'*. Royal patrons also seem to have been involved in the innovation of keeping ornamental carp in ponds, written records attesting that the Japanese emperor Keikou kept ornamental carp (which may have been local

**The goldfish is a common household and garden pet with an interesting history and economic value.**

carp species and not necessarily the progenitors of koi carp) in palace ponds in AD 94. Modern strains of nishikogoi, the Japanese word for what we in the West call simply koi or koi carp, originated in Japan in the 1820s, creating a global sensation when exhibited at a fair in Tokyo in 1914. Whilst koi are a pretty addition to a garden pond or indoor aquarium, to their devoted fans, the very best examples are 'living art' and as symbolic of status as a top-of-the-range sports car.

Many other British freshwater fish species grow too large for everyday indoor aquaria, though smaller specimens of tench, rudd and similar fishes do feature in some aquarist shops and also are kept in garden ponds, though often remaining elusive. Other fishes that have featured in indoor aquaria, though not so in recent times, include the spined loach. Like its near relative, the European weatherfish (*Misgurnus fossilis*, a fish that does not occur naturally in Britain), the spined loach can utilise atmospheric oxygen if that in the water becomes depleted, coming to the surface to exchange an air bubble in its mouth. It is this habit of greater activity under different climatic conditions that contributes to the common name of the weatherfish. Three-spined sticklebacks also make fascinating pets in unheated aquaria, small and hardy, the males becoming gaudy though highly territorial in the breeding season, and the species is willing to breed in captivity. However, as in many countries, native species seem to lack popularity amongst hobbyists regardless of their appeal to fish keepers in places where the species does not occur.

As we will see later, the introduction of exotic species into British waters through accidental or deliberate releases from the pet fish trade and some religious practices can give rise to serious consequences for aquatic ecosystems.

### 3.3.3 Fish and environmental protection

The early history of water quality management, as we will see in Section 11.4.1, initially focused on sources of pollution. This changed in Britain in the 1970s with the setting of river quality objectives (RQOs), framed not on effluent sources but on the water quality requirements of ecosystems in recipient water bodies. Various successive river quality classification schemes reflected the needs of different types of fish communities. At the highest quality level were river ecosystems supporting salmonid fisheries, at an intermediate quality level were different categories of coarse fish populations (mainly dominated by cyprinid fishes: members for the carp and minor family) and of lowest quality were 'poor' or 'bad' ecosystems, the latter supporting few or no fish. European legislation, particularly including the Freshwater Fish Directive (Council Directive 78/659/EEC of 18 July 1978 on the quality of fresh waters needing protection or improvement in order to support fish life: the first EC 'Environmental Directive' since revoked in 2013), echoed this transition towards the needs of aquatic ecosystems defined by the fish populations they supported. The quality requirements necessary to support fish populations were a surrogate for many other uses of fresh waters, the needs of freshwater fishes thereby serving as an important baseline for environmental management and human wellbeing.

Freshwater fishes thereby exert a profound effect on the economics of public water supply, both negatively and positively. Lakes and reservoirs used for water storage tend to become enriched by nutrient chemicals, particularly where poor agricultural practices or other forms of pollution occur upstream, stimulating the proliferation of planktonic algae that can further degrade water quality. The needs of fish can be a lever for more prescriptive controls on pollutant sources. However, as we will see later in this book, introductions of inappropriate fish species can compound these problems.

An interesting footnote here is the role that freshwater fishes are now playing in the fight against terrorism. Bluegill sunfish (*Lepomis macrochirus*), closely related to the pumpkinseed now naturalised widely in Europe and locally in southern Britain, is a hardy American fish commonly used as an experimental species in laboratories. The bluegill has now been deployed since 2006, in the post-9/11 era in the US, in terror warning systems in San Francisco, New York, Washington and some other major American cities. Bluegills are kept in small numbers in tanks flushed continuously by water abstracted for municipal water supply, serving as metaphorical 'canaries in a coal mine'. These fishes are sensitive to minute levels of toxins in water, far more instantly and acutely than laboratory test equipment. The breathing, heartbeat and swimming patterns of these test fish are monitored around the clock to indicate the presence of toxins, including 'coughing' reactions as they try to rid themselves of chemical toxins in the gills (though the fish are not susceptible to human-oriented biological agents). Computerised image monitoring triggers electronic warnings of possible contamination of the water supply. There

**The bluegill sunfish joins the fight against terrorism in the US. (Image © Jeffrey Divino.)**

are documented reports of at least one instance of the system intercepting a toxin before it got out of control and contaminated the public water supply serving New York City. The fish had responded to traces of (non-terrorist-related) pollution two hours earlier than it was picked up by other detection devices deployed by the city's Department of Environmental Protection.

### 3.3.4 Freshwater fishes and the written word

Very many great pieces of writing have revolved around fish, often also linked with fishing. The earliest known book written on fishing was *A Treatyse of Fysshynge wyth an Angle* in 1496 by Dame Juliana Berners. Dame Berners was an English noblewoman and prioress of the Sopwell Nunnery near Saint Albans and lived approximately between 1460 and 1500. Though perhaps not great literature, the book included many drawings of fishes with supporting text of great detail and vision, including essays on the virtues of conservation, the rights of streamside landowners and angling techniques and etiquette that embody today's basic ethics of sport fishing.

Dame Berners' work preceded the best-known example of early English angling literature, *The Compleat Angler*, written by Izaak Walton (1593–1683) and published in 1653, by approximately 150 years. *The Compleat Angler* is undoubtedly a work of great literature, written in the era of the English Civil War and evoking a bygone era of dew-drenched meadows, comely milkmaids and nut-brown ale in welcoming wayside taverns as part of an idyllic world that perhaps never

British freshwater fishes have been the subject of a wealth of books and other written media. (Image © Dr Mark Everard.)

really existed. However, *The Compleat Angler* was to become the second most reprinted book in English after the King James Bible. I have drawn from it liberally throughout this book, not merely to describe the fish but to celebrate the beauty of its language and the pastoral bliss it summons.

Other works of literate merit include the Reverend W. Houghton's 1879 *British Fresh-Water Fishes*, also referred to often in these pages. I also draw upon *The History and Topography of Ireland*, a valuable historical record published by the medieval clergyman and chronicler Giraldus Cambrensis ('Gerald of Wales' c. 1146–c. 1223).

Fish have stimulated literature too, as, for example, Henry Williamson's classic 1935 novel *Salar the Salmon* following the drama of its titular Atlantic salmon throughout its life. Other classic writers of angling and fishes include Hugh Tempest Sheringham (1876–1930), Bernard Venables (1907–2001) and 'BB' (Denys James Watkins-Pitchford MBE, 1905–1990). Great writing continues today, perhaps some of my own being of some merit but certainly that of my great friend John Bailey. The Medlar Press, based in Ellesmere, Shropshire, specialises in fish and angling books, both reissued and new, of high literate merit. The Medlar Press also published *Waterlog* magazine with the same high standard of literate quality, the first edition appearing in 1996 and running through to a 100th edition in 2017, after which it has reverted to a hardback periodic book format.

### 3.3.5 Freshwater fishes in music

'Die Forelle', German for 'The Trout', is the title of a famous piece of music by Austrian composer Franz Schubert (1797–1828). 'Die Forelle', Op. 32, D 550, set in the single key of D-flat major, is a lied (or song) composed in early 1817 for solo voice and piano, in which Schubert set the text of a poem by Christian Friedrich Daniel Schubart. The poem in its unedited version relates the tale of a trout caught by a fisherman, though the final stanza (omitted by Schubert) reveals that the tale of the fish's capture is a moral warning to young women to guard against young men. Schubert produced six editions of 'Die Forelle', each of them with minor variations. The song was popular with audiences of the day, with Schubert consequently commissioned to write chamber music based on the song, resulting in the *Trout Quintet* (D. 667) featuring variations on 'Die Forelle' in the fourth movement.

In contemporary music, many songs relate to fishing. Catfish feature as a common theme, particularly in blues genres. One of the theme tunes of the 2011 film *Bash Street*, directed by Ed Deedigan, is 'The Little Song of Little Fishes', inspired by Britain's small fish species and written and performed by one Dr Mark Everard.

### 3.3.6 Freshwater fishes in painting and other artworks

A particularly famous set of images of Britain's freshwater fishes was produced from the brushes of the British artist Alexander Francis Lydon, illustrator of the Reverend W. Houghton's

**The roach, as painted by British artist Alexander Francis Lydon as part of his series of illustrations in the Reverend W. Houghton's 1879 book** *British Fresh-Water Fishes*; **one of many classic paintings that have since been widely reproduced.**

1879 book *British Fresh-Water Fishes*. These are classic artworks in their own right, also used by the Portmeirion Potteries in their *Compleat Angler* range of tableware.

There are many great artworks featuring Britain's freshwater fishes, in formats from charcoal and pencil to woodcuts, paintings and many more and from the time of Dame Juliana Berners until the present day. Some of the highest calibre living or recent fish artists in whose work I find beauty and inspiration, some of which adorn my walls or bookshelves or illustrate the covers of my own books, include David Miller, Roger McPhail, John Searle, Bernard Venables, Mick Loates and Chris Turnbull.

Other paintings adorn once everyday objects, such as the charr dishes in which potted Arctic charr from the English Lake District, and Lake Windermere in particular, were packaged for their nationwide markets. These dishes were mainly produced in Staffordshire, but also Yorkshire, Scotland, Liverpool and Bristol. In a feature in the 100th edition of *Waterlog* magazine, Keith Harwood records that *'Zachariah Barnes (1743–1820), a potter in the Old Haymarket district of Liverpool, was one of the largest manufacturers of char pots and examples of his wares now command a high price'*. We have previously also noted the traditional, rather than a formerly commercial, practice on plumb lining fishing for Arctic charr in the deep waters of Lake Windermere as part of the culture of the English Lake District.

Fish have featured in sculptures, murals, wallpaper and many other artistic forms too numerous to mention, but pervasively so highlighting the ubiquity of their significance to us humans.

Fish taxidermy was formerly a significant feature of the decor of the avid angler's home, though in Britain it is more of a historic footnote in a more conservation-conscious era. In place of taxidermy, photographs, painted plaster and glass fibre casts, as well as carved wood facsimiles, have today largely replaced 'stuffed fish' as mementos of momentous catches. Nevertheless, mounted specimens are still prevalent amongst trophy anglers in the US, supporting a lucrative trade in taxidermy.

Our native freshwater fishes have also been enjoying a rising profile in television programmes such as BBC's *Springwatch* and *Planet Earth*.

### 3.3.7 Fish as symbols

Perhaps the best known of religious symbols associated with fish is the ichthys, derived from the Greek word for fish, consisting of two intersecting arcs resembling the profile of a fish. The ichthys is known colloquially as the 'Jesus fish', invoking the metaphor of 'fishers of men' amongst other meanings. The ichthys has featured in Christian symbology since the late second century, though, like many symbols and dates adopted by the church, its origins are pre-Christian, occurring in a number of other Near Eastern religions. The symbol of the fish is also symbolic of the vulva, and so of fertility and the many long-established traditions concerning it. Aside from the association of stew ponds with monasteries, there is however no strong affiliation of British freshwater fish species with religion in the same way that koi (cultivated strains of the common carp) have become important in some Buddhist gardens.

Fish are one of the groups of animals that feature in heraldry, usually associated with military families as symbols of bravery and steadfastness. The four main devices (symbols) in the *Fish* blazon (description in heraldic terminology) are the mullet, dolphin, fleur-de-lis and crescent, though more generally the species of fish is far from distinct. However, the pike is known

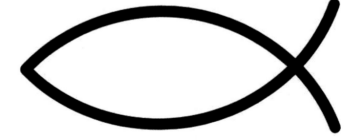

The ichthys, featuring in
Christian symbology since the
second century, though with
pre-Christian origins.

as the Lucy in English heraldry and the Ged in Scottish heraldry, usually blazoned as *naiant* (swimming), *embowed* (bowed) or *hauriant* (jumping). The eel also features in coats of arms, for example, of The Hague (the Netherlands).

Elsewhere in the world, fish can serve as spiritually or culturally significant symbols. For example, in Buddhism, fish symbolise happiness through their complete freedom of movement in the water. The golden mahseer (*Tor putitora*) has been identified as the 'Golden Fish', one of the eight auspicious symbols of Buddhism, symbolising the auspiciousness of all living beings in a state of fearlessness, without danger of drowning in the ocean of sufferings and migrating from place to place freely and spontaneously. Mahseer fishes have other significant cultural associations in South Asia, for example, appearing on the heraldic crest of India's former House of Mysore.

### 3.3.8 Fish and language

Given the significance of fish across multiple facets of human life, both here and across the world, it is far from surprising that fishy terms have infiltrated our sayings and common parlance. The term 'fishy' is just one obvious example, referring to something doubtful or of suspect motives, the origins of which are far from clear but may derive from the saying 'slippery as a fish' or the 'fishy' taste of meat as it is going off. Likewise, we use the terms 'minnows' or 'tiddlers' as a metaphor for small people and things. Then we have the term 'cold fish' to describe an unemotional person, and a 'queer fish' for someone who is eccentric or odd.

Other 'fishy' words may be coincidental. For example, the saying 'as sound as a roach', which is still prevalent in North West England, refers to someone or something that is perfectly sound. However, 'roach' in this instance is thought to be a corruption of the French word 'roche', meaning 'rock'. Likewise, it appears likely that the rudd and the chub derive their common names from the terms 'ruddy' and 'chubby', rather than the fish being the origin of the common adjectives.

There are also many commonplace sayings such as 'a pretty kettle of fish' to describe a muddle or awkward state of affairs, or 'a whole new kettle of fish' to describe an unexpected occurrence. These terms, it is believed, derive from a kiddle, or kiddle net, which is kind of a basket trap that was formerly set in a dam sluice in order to catch fish during Plantagenet times. Another explanation for this set of terms is that the 'kettle' was a local name for a cooking pot taken on picnics by Scottish gentry, used to cook salmon directly on the bankside for fishing parties. These sporting outings were known as 'a kettle of fish', one of the earliest recorded references to which is in Sir Walter Scott's 1823 book *St Ronan's Well*. It is not hard to envisage how mishaps could cause 'a pretty kettle of fish'. And so the enrichment of language goes on and on from ancient times to such modern uses as the Mafia term, popularised in *The Godfather* movies, that someone 'sleeps with the fishes' if the mob has killed them and dumped their body in water. How did we ever get by without such extensions to our language?

### 3.3.9 Fish twitching

The quest for more and bigger specimens may dominate the headlines of weekly angling publications. However, as we have seen, the reality is that the 'softer', qualitative values of the waterside experience are more widely appreciated by most everyday fishermen. The same is true of many waterside pursuits.

The immersive pleasures of exposure to wildlife associated with fresh waters are of course open to anyone with an affinity for aquatic environments, whether they go fishing or not. This includes enjoyment of all facets of wildlife, including the formerly much neglected fish fauna of our diverse rivers, ditches, ponds and lakes. For me personally, working on water systems globally, there is joy in catching or seeing new fish species regardless of size, as also in watching otters, water voles and other mammals, insects, trees and other vegetation, amphibians, birds, cloudscapes, the turn of seasons and in fact the whole waterscape experience.

Just as there are many botanists, butterfly enthusiasts and bird 'twitchers' in Britain, so too there is now a growing body of 'fish twitchers'. Appreciation of our rich freshwater fish fauna appears to be growing, if the increasing presence of freshwater fishes in wildlife broadcasting is a reliable guide. So, a whole chapter (Chapter 13) is dedicated to the topic of 'fish twitching'. May it bring you a lifetime of quiet, rewarding and perhaps formerly unsuspected pleasure!

'Fish twitching', observing fishes *in situ*, is an increasingly popular and rewarding pastime. (Image © Dr Mark Everard.)

## 3.4 HOW FRESHWATER FISHES USE PEOPLE

People benefit in so many ways from freshwater fishes. However, the benefits are also reciprocal, some fish species doing rather well at the hands of humans. With increasing global human mobility, a wide range of other species, from rats to cockroaches, organisms causing human diseases, or of sporting or ornamental use, have accompanied us across the world either deliberately or accidentally.

Brown trout and Atlantic salmon are native to the northern paleo-arctic, and rainbow trout naturally occur in the Pacific catchment of northern America, yet we humans have spread these species far and wide across the Northern and Southern Hemispheres as sources of food and sport. Brown trout, for example, are now widespread in Himalayan rivers, the Western Ghats of India, New Zealand and Australia, South Africa and Lesotho, North and South America, the Falkland Islands, Turkey, the Jordan River, Tanzania and Kenya, Japan, Pakistan, Swaziland, Fiji, Papua New Guinea, Morocco, Ethiopia, Cyprus and Nepal amongst many other regions where they did not naturally occur. Atlantic salmon escapees from fish farms have also formed viable breeding populations in river systems on the Pacific coast of North America, an ocean and continental divide away from their natural range.

**The brown trout has been widely spread by people across the world, arguably using humans to hitchhike into new environments.**

The common carp is another species, said to have been introduced to the UK and subsequently into Ireland by monks, that has been domesticated and spread for food and latterly for sport across a wide geographical range from cool temperate to tropical fresh waters in both the Northern and Southern Hemispheres.

Invasions by non-native species of fishes, plants, animals and microbes, though sometimes useful, cannot be assumed to be benign. Often, they have proved ecologically disastrous. As we will see in the following chapters, species such as common carp, topmouth gudgeon and sunbleak can radically change the nature and quality of aquatic ecosystems, fish populations and water resources.

In 2017, pink salmon, native to Pacific North America, made a confirmed appearance in British and Irish river systems, apparently as a result of straying and spreading from rivers draining to the Barents Sea into which they had been introduced by Russians. The longer-term consequences of the spread of pink salmon in British rivers are yet to unfold.

# KNOWING YOUR WAY ROUND A FISH

## 4

The medium of fresh water differs in many significant ways to that of air and land, so familiar to us and our terrestrially attuned senses and adaptations for mobility. And it is through adaptation to this altogether different medium that many of the characteristic features of fishes have evolved to enable them to go about the tricky challenges of breathing, swimming, feeding and breeding. To start with, fresh water is nearly eight hundred times denser and sixty-five times more viscous than air, properties offering both advantages and distinct challenges. One of the primary advantages of the higher density of water is that fish, water plants and all aquatic life are supported by the water because of the phenomenon of buoyancy. Fish benefit from life in a buoyant medium by a reduced need for a strong skeletal system to support their own weight. Neither do they have to expend substantial amounts of energy holding themselves upright or preventing themselves from sinking. Water also buffers temperature fluctuations far more than air, so fishes are protected from extremes of heat and cold and can consequently be harmed if exposed too long out of the water in sunshine or chilled air. However, the relatively higher density of water also creates higher viscosity, impeding the progress of a fish because of increased drag. Viscosity decreases inversely with water temperature such that water at 0°C is twice as viscous as that at 25°C.

## 4.1 FISHY SENSES

We comprehend the world around us through senses evolved as an adaptation to lives on land and immersed in air. The means by which fish sense their very different watery environment is therefore difficult to comprehend by us landlubbers, not only due to the adaptation of our senses but also the brain structures through which sensory information in processed.

### 4.1.1 Vision

Fish vision is adapted to the unique ways in which light moves through water. Water is a poorer conductor of light that air, compounded by its tendency to carry suspended or dissolved matter. Consequently, light travels relatively poorly through the impure waters that occur in rivers and lakes. Furthermore, water also tends to change the spectral composition of the light that it transmits, even over relatively short distances.

As a generalisation, predatory fish species tend to have the greatest visual acuity to aid location of prey by sight. Fish fry also tend to have relatively larger eyes than adults, in part as they tend to 'grow into their eyes' by body proportion as they age. The larger eyes of fry also reflect changing diet during the life cycle and the sense required to locate often mobile small animal food items and also probably to aid evading predators and assisting with shoaling behaviour. A bigger eye provides more receptor cells per visual angle, and in more sight-dependent adult fish, there is commensurate development of brain structures processing visual information. The zander possesses a *tapetum lucidum*, or reflective layer behind the retina, increasing its ability to see in dark conditions much like a cat's eye, aiding hunting in low light. Compared

The eye of a grayling.

to the general range of eye-to-body ratios encountered amongst the fishes, and in the size of the corresponding brain structure, generalist species such as roach do not show a high dependence upon vision. This is consistent with their largely non-predatory adult lifestyle, equipping them for life as generalist feeders.

The cone cells of the eyes of cyprinid fish are well developed, conferring them elaborate colour vision. Goldfish, for example, have been found to discern not only the spectrum of light visible to humans, but also extending into infrared and ultraviolet wavelengths invisible to us. Many species of fish with generalist, omnivorous life habits can survive when blinded, as, for example, can result from some parasitic infections or by physical damage, and many fishes are nocturnally active. Some fishes with tiny eyes, including catfishes and European eels, rely almost entirely on chemical and other senses to navigate, often nocturnally or in complete darkness. But vision is nonetheless often an important sense for fish as light travels exceedingly fast, visual systems enabling fish to make rapid responses to their environment. One of the limitations of eyesight in many fish species is that the eyes are positioned on either side of the head, providing a wide field of vision but not the locational faculty of stereo vision that humans and many other animals use to perceive distance. However, trout, amongst other fishes, have been found to be sensitive to the polarisation of light, which seems to help them discern distances to find food items and other objects.

### 4.1.2 The chemical senses

It is in the realm of the chemical senses that fish differ most profoundly from terrestrial animals such as human beings. We smell substances in the air and taste chemicals contained in matter in our mouths. But even for us, a great deal of what we describe as taste is, in reality, smell, as volatile substances enter the nose from the mouth cavity. This hazy borderline between smell

**The nostrils of a tench.**

and taste in air-breathing organisms is broken down almost completely for fish, the bodies of which are wholly and permanently bathed in water. Fish and other organisms living entirely underwater have the potential to literally smell and taste with their whole bodies. This enables them to detect gradients in chemical concentration within the water because of differences detected between the head and the tail. This turns the 'general chemical sense', a term for the combined taste/smell sense, into a directional means to detect, locate and home in on food or to move away from potential sources of danger.

In cyprinids, the taste buds are concentrated towards the head rather than towards the tail and are denser on the ventral (lower) than on the dorsal (back) surface, whilst the nocturnally active catfishes generally have dense chemical sensory organs all over their bodies. Taste buds in fish are responsive to tactile as well as chemical stimuli, effectively then enabling fish to taste, smell and feel food and other signals in the water simultaneously with the whole of their bodies. Compared to sound and light, chemical stimuli move only slowly through water, though they do persist so provide a different set of information such as concentration gradients in the water.

Chemical senses collectively are very important to most fishes for feeding, migration, reproduction and the evasion of predators. Consequently, the olfactory sense is remarkably acute, with some fishes found able to detect neutral amino acids at concentrations as low as $10^{-9}$ molar (this equates to $6.6 \times 10^{-8}$ g L$^{-1}$, about 2.5 ounces in 100 megalitres or, in more tangible terms, a water drop in an Olympic-sized swimming pool).

### 4.1.3 Pheromone detection
At least some fish species are also known to respond to a diverse range of pheromones (hormonal substances released into the water by their own and other species). Pheromonal

communication may be particularly important at night or in turbid water where visual clues are not available, and it appears important for a range of homing migration, shoaling, spawning and territorial behaviours. 'Aggregation pheromone' has been identified in roach, stimulating these social fish to crowd closer together. In contrast to the aggregation pheromones, 'crowding factors' have also been identified, and these substances stimulate the opposite reaction. It has been suggested that crowding factors may suppress reproduction or growth, contributing to stunting (populations of fish of reduced average size), thus preventing fish populations from outstripping available food supplies.

Minnows have been found able to detect and respond to the 'smell' of their predators. Predator pheromones which stimulate these reactions are known as 'kairomones'. This ability may be common to other fish species too. Recent research suggests that, in a piece of evolutionary 'chemical warfare', predatory pike adapt their behaviour by defaecating in areas remote from those in which they hunt. Another pheromonal response found in some fishes is the production of 'schreckstoff', also known as alarm substance, released from 'alarm substance cells' (ASCs) in the skin when the fish is physically damaged. A fright reaction is triggered in other fishes sensing this signal, resulting in changes in physiology and behaviour in order to reduce vulnerability to predation. The reaction may take the form of hiding, tighter shoaling, reduced activity, cessation of feeding, heightened awareness of danger and physiological stress reactions. For shoaling fish species, alarm signals are magnified throughout a shoal by visual, tactile and chemical clues.

The pheromonal system in cyprinid fishes has also been associated with learning. If, for example, minnows are exposed to schreckstoff at a particular place in an experimental aquarium, they will tend not to feed at that spot and to avoid it for a number of days. It is probable that if schreckstoff is regularly released at a particular place, fish can learn to associate that place with danger. The hormonal system of cyprinids therefore plays a key role in conditioning them to feeding, fright and flight, and reproduction.

### 4.1.4 The physical senses

The physical senses of fish are also, in many ways, quite different from our terrestrial experience. Owing to its higher density, pressure waves can travel nearly five times as fast through water than through air. Consequently, water is such an efficient conductor of sound that it is estimated that an underwater explosion in the oceans could be detected by a fish halfway around the globe. Many fishes have evolved efficient systems to detect pressure waves, both at frequencies audible to humans as well as many far below. Owing to the speed that water transmits sound, hearing enables rapid communication underwater. Unlike sight, hearing is not light dependent and is particularly well developed in species that are active at night. When some British freshwater fishes are taken out of the water, for example, chub, they may grate their pharyngeal teeth together to create sounds and vibrations. In some marine fishes, this has been found to be a means for communicating within shoals about potential danger, and this may be the case for fishes in general.

The difference in density between air and water has some interesting implications for hearing in fish. For fish species possessing a swim bladder, the ellipsoid air sac amplifies different resonance contained in sound waves. The inner ear is connected to the swim bladder wall, at least in most cyprinid fish species, and is extraordinarily sensitive to sounds at volumes and frequencies well below that discerned by humans. The connection between the inner ear and the swim bladder in body fishes is known as the Weberian apparatus, comprising a set of minute bones derived from the first few vertebrae physically connecting the inner ear to the swim bladder. The structure acts as an amplifier of sound waves. Sensitive hearing not only enables fish to identify different types of sound, but also provides information about its direction and distance.

Water pressure waves are also detected by the lateral line system, found in most groups of fishes. The lateral line comprises a series of sensory organs that is used to detect water movement, including vibrations and pressure gradients in the surrounding water. These sensory organs are known as hair cells, housed in pits in the flanks of the fish. These sensory hair cells lie beneath pores in the scales in fishes that are scaled. Sensors in the lateral line are adapted to detect low-frequency sounds, surface waves, currents and alterations in pressure. The lateral line serves many important roles, including detecting currents, reflections and disturbances, with roles in shoaling, predation detection and orientation. Even experimentally blinded minnows are able to navigate quite adequately by utilising the effects of current upon the objects around them and by the reflection of pressure waves generated by their own body movements. Sensitivity to water pressure varies amongst fish species according to their adapted life habits and ecological needs, the day-active perch that hunts substantially by sight having a less developed lateral line than species remaining or predominantly active by night. These senses appear to be important for a range of behaviours, including in mating and aggression. This capability, in broad terms, is similar to the echolocation system in bats (albeit that bats tend to use far higher ultrasonic frequencies), enabling some specialised cave-dwelling freshwater fish species to live their whole lives in complete darkness. Though no species specifically evolved to live in caves is found in Britain, a number of species such as brown trout and bullheads are found in underground streams. In many fishes, the lateral line is visible as a complete or partial line mid-flank running from behind the operculum (gill cover) to the tail, though it is reduced in some fish species. The extent of the lateral line can be an important feature used in fish identification.

Tactile, or touch, receptors are widespread both on the surface and within the tissues of fishes and play important roles in shoaling and spawning behaviours. Proprioceptors, or stretch

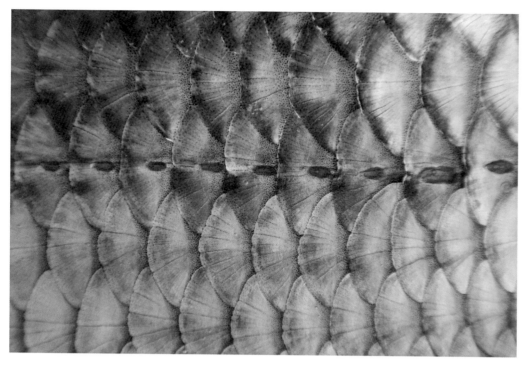

The lateral line of a roach. (Image © Dr Mark Everard.)

receptors, are also embedded in the body tissues, both regulating swimming activity as well as governing buoyancy as gas in the swim bladder expands under reduced pressure as the fish rises in the water column. Thermal, or temperature, sensors also play important roles in controlling the general metabolism of fishes.

### 4.1.5 The electromagnetic senses

Many fishes also possess electrical sensors, though their presence in British freshwater species is largely unknown. This sense is best known in sharks, using specialised 'ampullae of Lorenzini' to track down prey by the electrical signals that they give off. However, given the comparatively high conductivity of the water environment, many fish biologists believe that this ability may be present to some extent in all fishes.

There is also some evidence to suggest that goldfish, and perhaps other fishes too, may have the ability of magnetic orientation, which may play roles in homing migrations prior to spawning.

### 4.1.6 Fishy perception

How then do freshwater fishes perceive their world? These diverse senses combine to provide a wealth of signals about the environment. These signals are all processed by a brain incapable of the human experience of 'consciousness' because of the lack of a neocortex (massive development of the frontal part of the brain) as found in humans and many higher mammals. Rather, the responses of fishes are predominantly reactions to stimuli either evolved of learned by experience, rather than consciously computed.

At this reactive level, and based on the relative acuities of the senses and the degree of development of the corresponding brain structures, fishes can be separated into three main specialised groups: (1) visual fish, which tend to be predatory; (2) tasters, which tend to be benthic (river or pond bed) feeders and (3) acoustico-lateralis, which rely upon their sound/ pressure systems and are largely pelagic (open water) species. Dependency on these senses can change between life stages even in the same species. For example, the fry of roach, as indeed many species, are principally visual creatures, reflecting their need to hunt zooplankton (small invertebrate animals in the water column) and other small benthic invertebrates to sustain a rapid growth rate. However, as the fish grow, their sensory systems develop to best fit the ecological niche to which the species is adapted. The eyes of perch remain large, enabling them effectively to hunt prey by sight as adult fish. The initially large eyes of young common bream species do not develop at a rapid rate, resulting in adult fish with reduced reliance on vision but enhanced taste systems, suiting them to life browsing on invertebrate prey buried in soft sediments.

## 4.2 BODY SHAPE

The overall body shape of a fish tells us much about the environment to which it is adapted, in terms both of the way that this propulsive power is used and the water conditions and life habits to which it is adapted. Species such as common bream are deep-bodied and strongly laterally compressed as an adaptation to stable swimming in still or slow-moving waters. This stable body form – larger common bream are sometimes nicknamed 'dustbin lids' – enables these fish to orient themselves intricately to feed on small food items buried in muddy beds. By contrast, trout species and dace are streamlined in profile and have a rounded cross section, adapting them to life in faster flows in open water. Barbel are also streamlined with a torpedo-like body profile but are notably flattened on the underside, suiting them to life on the bed of swift rivers.

Behind the head, and surrounding the cavity of essential organs towards the front of the body behind the gills, the rest of the flanks comprise striated muscle blocks that typically account for between 40% and 60% of the mass of a fish. This musculature, anchored to the backbone and the finer bones radiating upwards to the back and downwards to the ventral surface, is crucial for locomotion.

Pike are elongated and streamlined, perfectly adapted to their life as an ambush predator. The extensive flank muscles of the pike generate considerable power for rapid acceleration to

Body shapes differ radically amongst fish species, as illustrated here by the high-backed and laterally compressed common bream, the streamlined dace, the barbel flattened beneath for life on or close to the river bed and the snake-like European eel.

intercept prey. Another elongated, sinuous body form is seen in the eel and species of lamprey, adapted to slow swimming, accessing tangled vegetation and lurking in soft sediment.

## 4.3 FIN CHARACTERISTICS

As important as body shape to lifestyle is a means of propulsion through the viscous medium of water. This is the purpose served by the fins. Even 'fish' that are not true fish, such as the lampreys that stem from a far more ancient evolutionary line than modern 'bony fish', have flattened fin-like vanes on their back, rear and underside to aid swimming.

Most other fishes of Britain's fresh waters are of a more recent evolutionary lineage, part of the order of 'bony fishes' known as the teleosts. The teleosts have a number of key features that explain their success in adapting to a bewildering array of habitats and diets, particularly including adaptable mouth parts, thin flexible scales and rayed fins. It is the rayed fins that give the teleosts their common name: the 'ray-finned fishes'. More ancient fossil fishes had spines supporting skin flaps, but replacement of spines with soft rays to support the fins gave teleost fishes far finer control over their direction, orientation and propulsion. Fin size and placement

**The dorsal fin of a roach.**

give a strong indication of the life habits of the fish, as well as being important features in the accurate identification of species.

There are basically two types of fins: the 'median fins' and the 'paired fins'. Median fins, on the midline and perpendicular to the body, act as planes stabilising the fish on its midline, ensuring that it is able to swim straight without losing balance, as well as enabling rapid forward propulsion. Median fins include the dorsal (back) fin or fins, the caudal (tail) fin at the rear and the anal fin beneath the fish behind the anus. Median fins are usually well developed in fast-swimming fish, particularly predators, enabling them to convert the power of their body muscles into forward motion. By contrast, small, paired fins are a later evolutionary development, providing lateral stability. Paired fins behind the gill (pectoral fins) and on the underside (ventral fins) are commonly better-developed in slower-swimming fishes, providing fine control over orientation, balance and slow, paddling swimming. The combination of median and paired fins provides fish with dexterity, power and swimming control, so the position, size and structure of the fins reflects the mode of life to which fish species are adapted.

Some extreme examples here are the pike, on which the dorsal and anal fins are set well back on the body near the tail fin, providing this ambush predator with explosive acceleration. By contrast, tench are omnivorous fishes of still or sluggish waters, with smaller dorsal and anal fins further towards the middle of the back and underside to maximise lateral stability, but with well-developed paired pectoral and ventral fins offering more precise control for slow, gentle swimming as these fish forage for edible items on the bed of still or slowly flowing waters, particularly at night.

The leading edge of soft fins is commonly supported by a cluster of fused spines in many British freshwater fish species, with soft, branched rays supporting the rest of the fin. Some

British freshwater fishes have a single dorsal fin, whilst several others have two. One species, the burbot (now declared extinct in Britain), has three dorsal fins. If there are two dorsal fins, at least for British freshwater species, the first is held erect by hard spines. The two dorsal fins may run into each other (as in the ruffe) or be separated by a clear gap (as in the closely related perch). Scientifically, the spines are counted in Roman numerals, whereas soft rays are counted in Arabic numerals. For example, the spine and ray count of the dorsal fin of the roach is III/9. The spine and ray count of the dorsal and anal fins is a particularly helpful visual feature used when distinguishing otherwise broadly similar fish species. So too may be the orientation of the fins with respect to each other.

## 4.4 SCALES, THE LATERAL LINE AND MUCUS

The bodies of most species of British freshwater fishes are covered in scales. The scales are bony, overlapping plates providing both armour and flexibility. The evolution of thin, flexible scales amongst the teleosts offered them significant advantages compared with older groups of fishes from the fossil record that were heavily laden with bony armour or thick, bony scales. By contrast, the thin, bony scales offer greater flexibility, contributing to a lighter weight and more efficient swimming.

Fish scales may be large and prominent, fine and less obvious, apparently absent or may be completely lacking as in the case of bullheads, stone loach, American black bullhead and some cultivated forms of the common carp. Three-spined sticklebacks exhibit a variety of scale patterns across their wide geographical range, some strains with scales reduced to a few large, bony armour plates known a scutes. The common sturgeon also has a row of scutes along its otherwise scaleless flanks. The presence or absence and the number of scales varies between fish families and species and can be helpful in distinguishing amongst similar fish species. This

**Scutes, or large lateral scales, on the flanks of a common sturgeon.**

is particularly true of the number of scales along the lateral line. Furthermore, the lateral line may be complete, running as far as the tail, or else may stop short.

As noted when considering fish senses, the lateral line comprises a series of sensory organs used to detect water movement, vibrations and pressure gradients, comprising pits in the flanks of the fish visible as a line of pores in scales on fishes that possess them.

The body surface of fish is also covered in a thick, slimy lubricating material known as mucus, manufactured by specialised mucus glands widespread in the skin. The thin layer of skin producing this mucus lies above the layer of scales. Skin and the mucus layer play crucial roles in protection against infection, shedding of parasites, and aiding swimming efficiency by reducing friction, turbulence and cavitation as water flows over the smooth and softened surface of the fish.

## 4.5 MOUTH, BARBELS AND TEETH

The orientation, size and other features of the mouth also reveal a great deal about the life habits of fish species. Bleak and rudd, for example, have mouths that are 'superior', or in other words turned upwards, better suiting them to feeding from or near the water's surface. By contrast, the mouths of barbel, gudgeon and loach species are 'inferior', or angled downwards, equipping them for life on the river bed. Other fish, such as chub, dace and roach, have mouths positioned centrally, enabling them to readily exploit food items across a range of depths.

Perhaps unsurprisingly, some predatory fishes, such as pike and zander, have particularly strong teeth in their jaws to enable them to hunt fish and other live prey. The jaw teeth are smaller but nonetheless stout in the predatory Atlantic salmon and brown trout. Other predatory fishes, such as perch but also zander and pike, have an array of smaller, backward-pointing vomerine teeth in the roof of the mouth, preventing prey engulfed in the mouth from escaping.

The mouths of some fish species are surrounded by pairs of fleshy, chemically sensitive barbels, or 'whiskers', that help the fish search out food. Gudgeon and tench, for example, have a single, small barbel either side of the month. Barbel have two strong pairs of barbels, giving the fish their common name (and across the world many catfish species are also known locally as 'barbel' because of their prominent 'whiskers'). Stone loach and spined loach have three pairs of barbels. By contrast, many other species such as chub, roach, ruffe, European eels and crucian carp lack barbels entirely.

The mouths of different cyprinid fishes reflect their life habit and diet, as, for example, the upward-pointing mouth of the rudd, the centrally oriented mouth of the chub and the downward-oriented mouth of the juvenile barbel.

An array of backward-pointing vomerine teeth on the roof of the mouth of this pike enable it to hold onto prey that it has captured. (Image © Dr Mark Everard.)

Other fish species, including all members of the carp and minnow family (the cyprinidae or cyprinids), lack teeth in their jaws but instead have pharyngeal teeth (powerful teeth deep in their throat modified from gill rakers) used to crush hard food items including shells and seeds. The structure of the pharyngeal teeth is an important feature used to definitively identify closely related and hybridised fishes. However, as it is necessary to kill and dissect the fish to see them, analysis of pharyngeal teeth is a job for specialists only.

## 4.6 THE SWIM BLADDER AND BUOYANCY

Many British freshwater fishes possess an internal air sac known as a swim bladder within the body to control their buoyancy. Some fishes, such as the European eel and loach species, lack a swim bladder.

Perch and similar 'advanced fishes' possess a swim bladder that is sealed, and is therefore inflated and deflated with pneumatic glands. However, fishes of the carp and minnow family (the cyprinids are the largest family of freshwater fishes globally including in British waters) lack these glands, 'topping up' the swim bladder by gulping air from the surface to increase buoyancy, or releasing bubbles if the fish needs to become less buoyant. Consequently, species such as common bream, roach and tench may be seen 'priming' at the surface, often at dusk, to gulp in a bubble of air as a prelude to moving into shallower, food-rich waters where they may have been more vulnerable to predation during daylight hours.

## 4.7 COLOUR

In the descriptions of fish species in Chapter 5, the colours of their bodies, fins and eyes are mentioned as a general guide. However, colour can be a poor or misleading characteristic for definitive identification. In the dark, or in murky water, many fish species can become quite pale and can appear so relatively quickly, for example, when put into a white bucket. In summer, generally silvery fishes such as roach feeding extensively on insects and crustaceans can become distinctly golden on the flanks with vivid scarlet fins as carotenoid pigments in their invertebrate diet accumulate on the body and fins. This may give some roach at this late-summer time of year a superficial resemblance to the generally gold-tinged rudd. Conversely, rudd in cloudy water, feeding sparingly on invertebrates or under stress can take on a washed-out silvery hue superficially resembling roach. Though tench generally appear olive green (though may be brownish) and perch usually have bold black vertical stripes over a greenish body, even these strong colours may be modified by environmental conditions. Colour is therefore far from the most reliable characteristic for definitive identification of species.

## 5

# THE PISCATORIAL *DRAMATIS PERSONAE*

Our prior discussions of the things that freshwater fishes have done for us, and of their distinctive features, have been illuminated by examples of a range of freshwater fish species. So let's now take a more structured approach to the diverse species of fish that inhabit the varied fresh waters of the British Isles.

Many books already provide exhaustive lists of our fresh water fish fauna, together with their distinctive and comparative features, with notes on their biology and ecology. I am guilty of writing a few myself, with some of my books listed in the 'Further Reading' section. Many of these books run through alphabetically by family names beginning with the sturgeons (Acipenseridae), the British species of which is so rare and elusive that few wildlife watchers let alone fish specialists have ever seen or could ever hope to see one in the wild. But this is not that sort of a book; rather, it is an 'upside down book on fishes'. It is instead primarily concerned with interesting things about the British freshwater fishes as important elements of the wider wildlife and ecosystems of these islands and their place in our culture, rather than yet another simple identification guide.

The purpose of this chapter is to introduce in relatively brief terms the species found in Britain, the families to which they belong and some other relevant details of their habitat preferences and mode of life. It does so in much the same way that a playwright might list the *dramatis personae* in the front matter of their play, so that we can refer back to recall who's who when we meet them in the thick of the action. So here we go, but not in alphabetical order. In the spirit of this book, it is the most common and widespread fish families and species that we will encounter first.

The distribution of these fishes across Britain is mapped in Appendix 2.

## 5.1 FISHES OF THE CARP AND MINNOW FAMILY

The fishes of the carp and minnow family (Cyprinidae, generally commonly referred to as the cyprinids) are some of the most widespread and commonly encountered of British freshwater fishes. The cyprinids form the largest family of freshwater fish globally, with over 2,400 species in about 220 genera across North America, Africa and Eurasia. They are also the best-represented family of British freshwater fishes, including nineteen species, twelve of which are native and seven introduced. All cyprinids lack stomachs and have toothless jaws. Their bodies are covered evenly by scales that may be conspicuous or small (with the exception of some scaleless, artificially reared strains) and have a single dorsal fin supported by soft rays behind fused spines at the leading edge. Though the jaws are toothless, the cyprinids possess pharyngeal (throat) teeth formed from modified gill rakers. The pharyngeal teeth are used to crush food items as they are swallowed and can be used by specialists to confirm the identity of closely related species and also hybrids among them.

So large and diverse is the family Cyprinidae that there is evolving consensus it spans a cluster of families within an order of Cypriniforme fishes. This is not the place to go into the minutiae of fish taxonomy, particularly as consensus is still developing amongst scientists. Suffice to say that one of the major divisions is a group of Holarctic (Northern Hemisphere) cyprinids formerly known as subfamily Leuciscinae, or dace-like fishes, that may be classified as a separate subfamily Leuciscidae. Moving on from the technical details, we will consider these dace-like fishes first, as they are amongst the most commonly encountered of Britain's freshwater fish fauna.

The roach (*Rutilus rutilius*), often known as the redfin, is one of the most widespread species in British waters, with an adaptable life history that sees it as at home in a mighty salmon river as in a small duck pond. Though not native to Ireland or much of Scotland, west Wales and the far southwest of England, roach have become well established in these places following introductions. Roach have a deep, laterally compressed body covered evenly by large, conspicuous scales. Body colour is generally silver, darker on the back and pale beneath, with bright red fins. The mouth lacks teeth and barbels. Roach are omnivorous, juveniles feeding extensively on small invertebrates and algae and older fish tending towards a more plant- and detritus-based diet, though also favouring molluscs (freshwater snails and small mussel species). Roach spawn communally on submerged vegetation in the spring, the sticky eggs receiving no subsequent parental care. Roach can be found in powerful rivers, small streams, canals and ponds as well as large lakes and estuaries, reflecting their adaptability to a range of conditions and diets.

The rudd (*Scardinius erythrophthalmus*) is superficially similar to the roach, but is distinguished most obviously by the leading edge of the dorsal fin being set back two to three scale columns from that of the ventral fins (those of the roach are in line). Rudd also have an upturned mouth,

The roach.

**The rudd.**

whilst that of the roach has a clear overbite. Rudd have a deep body profile with an even covering of large, conspicuous scales over a generally golden-silver body colour and crimson fins. The mouth lacks teeth and barbels. Rudd occur in some slower rivers, but are primarily fish of still waters. They often occupy middle and upper layers of the water column. Mature rudd in food-rich waters can take on a vivid golden and crimson colouration. Rudd spawn communally on submerged vegetation in the spring, the sticky eggs receiving no subsequent parental care. Young rudd feed extensively on small invertebrates, becoming increasingly omnivorous as they grow.

Chub (*Squalius cephalus*), also known as chevin and loggerheads, are a more robust fish of flowing waters, though they can prosper but not breed in still waters. The body profile is streamlined and rounded in cross section, but notable 'chubby', with an even covering of large, brassy and conspicuous scales. The dorsal and tail fins are dark and the ventral fins reddish. The mouth is large, lacking teeth and barbels. Chub spawn in shoals on river gravels in late spring or early summer, with no parental care once eggs are laid. The fry require warm, shallow marginal water to enable them to grow quickly enough to withstand winter spates. Cool summers with poor growing conditions often result in the loss of most of the season's juveniles by year end. Consequently, strong and weak year classes are found in chub populations, tracking reproductive success in optimal prior summer nursery conditions. Juveniles feed on algae and small invertebrates. Older fish become increasingly omnivorous as they grow: plants, fish, amphibians and other small animals all make up the diet.

Dace (*Leuciscus leusciscus*), also known locally as dare or dart, superficially resemble small chub from which they are distinguished by a more delicate mouth and concave outer borders to the dorsal and anal fins (those of chub are convex). Dace occur throughout mainland Britain, except

The chub.

the far north and west. They are lively fish of rivers, though, like chub, they can prosper though not breed when introduced to still waters. Dace have slender, streamlined bodies covered evenly in large, conspicuous scales. Body colour is generally silvery, darker on the back and pale or white beneath, with pale fins. The mouth lacks teeth and barbels. Dace are common in flowing waters, particularly those of good quality, across the British Isles and much of northern Europe. They are active at all depths of the water column, including taking food items from the surface, particularly during warmer months. Dace spawn communally on river gravels in late winter or early spring, one of the earliest spawners amongst the British freshwater fish fauna, depositing sticky eggs that receive no parental care. Juveniles exploit the spring boom of available food, adaptation to cold conditions meaning that dace continue to feed and grow throughout the year. Young dace require food-rich warm water in river margins to maximize growth during their first summer. Adult fish remain omnivorous, but preferentially feed on larger invertebrates and fish fry.

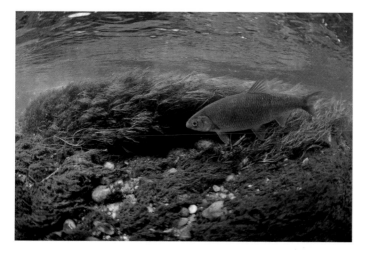

The dace.

Orfe (*Leuciscus idus*), also known as ide, are better known in their ornamental form, the golden orfe, which is a common garden pond and larger cold water aquarium fish. It is by this route that this species, native to many parts of continental Europe and Asia, has escaped to colonise a number of British rivers predominantly in the south of the country. Naturalised stock reverts to the wild silvery colouration, paler below and darker on the back. Sometimes confused with dace or chub by those unfamiliar with the wild strain of the species, orfe have a far denser covering of relatively small but conspicuous scales. The outer edges of the dorsal and anal fins are convex, the tail is forked, and the mouth lacks teeth and barbels. In their native range, orfe migrate up tributary streams to spawn communally on gravel or submerged vegetation in early spring, the sticky eggs receiving no parental care. Juvenile orfe feed on algae and small invertebrates and tend to be gregarious. Larger fish progress to a diet comprising mainly insect larvae, other invertebrates and, opportunistically, fish fry.

The common bream (*Abramis brama*), generally referred to as bronze bream or simply bream, can grow large (up to 9 kg) and are a fish of slow rivers and still waters in lowlands and more nutrient-rich areas of the British Isles. Common bream have high backs and strongly laterally compressed flanks, adapting them to still and slow-moving waters. The body is covered in large scales, and the mouth lacks teeth and barbels but can be protruded to help the fish suck up food from river or pool beds. The common bream is a widespread shoaling species found in still and slow-flowing waters, particularly in the south and east of the country. Their diet is omnivorous, the high back and compressed sides providing a great deal of stability adapting them to grub around for invertebrates and other food in soft sediment. Common bream spawn communally in the late spring or early summer, shedding eggs on submerged vegetation and exhibiting no parental care. Juvenile common bream feed extensively on small invertebrates, largely on bloodworms (chironomid midge larvae) and other small organisms, older fish tending

to have a more varied diet feeding primarily on the bed of the river or lake. The body colour changes with age, silvery in younger fish but darkening to bronze with age.

Silver bream (*Blicca bjoerkna*), also known as white bream or bream flat, are also laterally compressed, inhabiting slower rivers and still waters in lowland southern England. Their deep, laterally compressed bodies are covered evenly with conspicuous scales, and the fins are colourless or slightly reddish. The small head has a moderate hump (occiput) behind where it joins the body, and the mouth lacks teeth and barbels. Silver bream are easily confused with common bream, but have fewer scales along the lateral line, a mouth that does not protrude into a tube, and a far larger eye that is typically a quarter the diameter of the head. Silver bream are typically found near the bed where they shoal and feed. Silver bream spawn communally during the late spring or early summer on submerged vegetation, often in several discrete spawning events, releasing sticky eggs and exhibiting no parental care. Juveniles grow rapidly, feeding extensively on small invertebrates, whilst older fish tend to have a more omnivorous diet.

Bleak (*Alburnus alburnus*), sometimes known evocatively as the river swallow for their habit of vanishing in winter and reappearing in large numbers in the spring and summer, are a small native species of rivers, adapted to feeding in and near the surface film and often forming dense shoals. Bleak have prominent scales, generally silvery in colour, but are darker and commonly greenish on the back. The body is streamlined and laterally compressed. The large mouth, lacking teeth or barbels, as well as the eyes, are upwardly pointing, reflecting the species' surface-feeding habit. The fins are colourless, and the base of the anal fin is particularly long. Bleak spawn during the spring on stones, gravel or nearby vegetation in shallow water, the sticky eggs receiving no parental care. Juveniles feed extensively on

**The silver bream.**

small invertebrates and algae, whereas older fish feed opportunistically on small food items, particularly invertebrates, borne on the current or the water's surface.

As noted, the preceding species – roach, rudd, chub, dace, orfe, common bream, silver bream and bleak – belong to a branch of the carp and minnow family that characterises them as 'leuciscine' (or dace-like) species. Given their relative genetic closeness, hybrids are possible amongst leuciscine species. Some occur rarely because of different spawning habitat requirements or timings across the seasons. However, where roach, rudd and bream species (in particular) occur in habitat-poor waters, all of them competing for common spawning sites on vegetation during the late springtime, hybrids are common. As has been noted in

**The bleak.**

THE COMPLEX LIVES OF BRITISH FRESHWATER FISHES

**A rudd x bream hybrid.**

consideration of the general characteristics of fishes, and of the species listed above, the number of spines and soft rays in the dorsal and anal fins and the relative fin positions can be important for distinguishing between similar species and their hybrids. A table documenting the tendency of leuciscine species to hybridise is included in Appendix 3. Appendix 4 tabulates some of the more obvious external identification features of the cyprinid fishes. Much more could be said about leuciscine and other hybrids, but suffice to say that we will return to relevant aspects when considering habitat and reproduction later in this book.

Next we come to the carps. The largest of these is the common carp (*Cyprinus carpio*). There are many forms produced by aquaculture and the aquarist trade including king carp, koi, leather carp and mirror carp. The common carp is a large fish, growing to 27 kg and more in Britain and larger still internationally. It has a conspicuous covering of large scales with a brassy colour, though cultivated forms may lack scales and be differently coloured. The dorsal fin is single and elongated, and two pairs of short barbels surround a toothless mouth that may be protruded to aid feeding. Common carp are an introduced species, brought to the British Isles here in successive waves from at least the fifteenth century because of their hardiness, omnivorous habits, rapid growth and consequent uses for aquaculture and food, sporting purposes and as ornamental fish. The voracious, omnivorous diet, fast growth and large size and habit of grubbing up sediment can make common carp problematic through disrupting aquatic ecosystems, often profoundly. Common carp spawn communally over dense vegetation from late spring to summer when the water reaches 18°C–20°C. Male fish chase and jostle gravid females, which shed small, sticky eggs that adhere to submerged vegetation, subsequently receiving no parental care. Common carp are omnivorous with voracious appetites throughout their life cycle. The common carp was reputed to have been brought to our isles in the late fifteenth or early sixteenth century, introduced by monks as a source of food; in *The Compleat*

*Angler*, published 1653, Izaak Walton's Piscator notes that the 'queen of rivers' was now naturalised but that,

> *...doubtless there was a time, about a hundred or a few more years ago, when there were no Carps in England, as may seem to be affirmed by Sir Richard Baker, in whose Chronicle you may find these verses*:
>
> *Hops and turkies, carps and beer, Came into England all in a year.*

In his 1817 book *Elements of the Natural History of the Animal Kingdom*, Charles Stewart is rather more exact (though corroborating sources are not known) stating that the common carp '...*was introduced into England in 1514, and into Denmark in 1560*...' It has been postulated that the common carp was introduced to Europe from Asia via Cyprus, giving rise to the Latin name *Cyprinus carpio*, though this remains hard or impossible to prove given the widespread introduction and subsequent rapid spread of the species.

Crucian carp (*Carassius carassius*), also known as crucians or carassin (French), are a more delicate carp (growing only to a maximum of 3–4 kg) that tend to inhabit well-vegetated still waters. Crucian carp do not compete well with larger, more robust species and in open waters. Instead, they are tolerant of low oxygen conditions and both high temperatures and prolonged ice cover and can thrive in densely vegetated pools and marshes where other species fare only poorly. Crucian carp have notably humped backs, and are evenly covered with conspicuous scales over a brassy or olive-coloured body. The mouth is small and lacks barbels. The dorsal fin is long with a strong, slightly serrated third spine preceding the soft rays. Initially native to the eastern side of England, crucian carp are now more widespread across the British Isles through introductions. Crucian carp spawn in the late spring through to the summer, females shedding small, sticky eggs into dense vegetation in shallow water. There is no subsequent

**The crucian carp.**

parental care. Juvenile and adult crucian carp feed on small animals and plant matter throughout their life.

Goldfish (*Carassius auratus*), also known as brown goldfish when reverting to a natural bronze body colour in the wild, are scattered throughout the British Isles by introduction into a wide variety of water bodies through escapes and releases from garden ponds and aquaria. They are now widely naturalised throughout Britain. Superficially, goldfish resemble crucian carp and are often confused even by experts. These small, hardy carps are native to the still waters of central Asia, China and Japan, tolerating poor water quality, and are well adapted for life in small ponds or thickly vegetated pools and marshes. Goldfish spawn communally when the water temperature exceeds 20°C, females releasing small, sticky eggs that adhere to water plants and subsequently receive no parental care. Juvenile and adult fish feed on small animals and plant matter throughout life.

**The goldfish, in its natural colour.**

The grass carp (*Ctenopharyngodon idella*), also known as the white Amur, is a larger carp (growing to 45 kg) native to eastern Asia from China to eastern Siberia in still waters, large rivers and brackish waters across a wide temperature range (0°C–38°C). However, it has been widely stocked around the world including locally in still waters throughout the British Isles. Grass carp have elongated, cylindrical bodies evenly covered with prominent scales of an overall bronze to olive colour that fades to paler beneath. The snout is short and the mouth is terminal, lacking barbels or teeth. The species was initially introduced to the British Isles to control aquatic weeds, as it has a pronounced herbivorous diet. As grass carp are not known to breed in British waters; the species persists here solely through stocking, which today is mainly for ornamental and angling purposes. Grass carp spawn in their native regions on river beds with very strong currents and at relatively high temperatures of 20°C–30°C, the eggs drifting in the current with no parental care. The requirement for high temperatures and long rivers are factors leading to the failure of grass carp to breed in British rivers.

The first of these three carps – common carp, crucian carp and goldfish – tend to interbreed, with hybrids possessing intermediate features. Hybrids of goldfish and crucian carp with common carp are usually betrayed by the presence of small barbels around the mouth. Hybrids between crucian carp and goldfish are very difficult to determine. The widespread introduction of goldfish is threatening wild stocks of crucian carp with extinction through hybridisation and competition. The grass carp is more remotely related to these other three carps, and also has different breeding habits so is consequently not known to hybridise with them.

Some other foreign carp species have also rarely appeared in British waters, though are not considered here as part of the established fish fauna. These include, for example, blue carp (*Mylopharyngodon piceus*), also known as black carp or snail carp, native to China though

**The grass carp.**

rarely encountered in British fresh waters as a survivor of historic stocking as an ornamental fish. More concerning is the silver carp, also from China, a major global invasive species, some remnant stockings of which occasionally appear but which should be removed under fishery protection legislation.

The barbel (*Barbus barbus*), often referred to as beard or whiskers (particularly by anglers), is another large cyprinid (growing as large as 9 kg). A powerful and streamlined fish, barbel naturally occur in rivers draining to the North Sea from the Humber to the Thames systems, but are now far more widely introduced across England and some of Wales. Barbel have streamlined bodies evenly covered in conspicuous bronze or olive scales, with strong, amber fins – the pectoral fins in particular are well developed – adapted for life on river beds. The mouth is underslung, lacking teeth but with strong, fleshy lips and two pairs of barbels used to locate food. These fish inhabit the lower layers of water where they feed on the river bed as opportunistic omnivores. Barbel are a shoaling species that Izaak Walton describes will '...*flock together like sheep*'. Barbel spawn communally over well-flushed gravels during the late spring or early summer, often disturbing the gravel bed into which their sticky eggs may fall offering a little protection, though there is no parental care. Juvenile barbel feed extensively on small invertebrates, though as barbel grow, they adopt a more omnivorous diet that includes any animal and plant matter that they can root out from river beds using their sensitive barbels and strong, fleshy lips. As the British Isles are at the extreme northern extent of the barbel's European range, spawning does not automatically occur every year, and neither is fry survival assured, particularly when cool summers inhibit their growth and consequent ability to withstand autumnal spates.

Gudgeon (*Gobio gobio*), also affectionately known as gobies or gonks, are a smaller fish (up to 150 g) and are widespread in rivers and some still waters. They are an attractive, shoaling fish

**The barbel.**

**The gudgeon.**

with prominent, pearlescent scales, the flanks mottled with an iridescent overlay. The body is streamlined and flattened beneath, reflecting adaptation to life on the river bed. The fins are colourless but have some dark mottling. The underslung mouth lacks teeth, and a small, single barbel is present at each corner (distinguishing gudgeon from small barbel which possess two pairs of stouter barbels). Gudgeon are widely distributed throughout mainland Britain, tending to shoal. Gudgeon spawn communally in shallow water on gravel, sand or vegetation from late spring to early summer, usually when water temperatures approach 15°C, and they do so at intervals over several days. Gudgeon eggs stick to submerged surfaces but receive no subsequent parental care. Juvenile gudgeon feed extensively on small invertebrates and algae, older fish feeding opportunistically on the river bed on a range of small plants, animals and other organic matter.

Minnows (*Phoxinus phoxinus*), also known as penk, are another small (growing up to 20 g), shoaling fish. Minnows commonly occur in rivers, but also inhabit some larger, well-oxygenated still waters across mainland Britain. Minnows can be profuse in suitable waters, familiar as the archetypal 'tiddler' we might have pursued in brooks and streams as children. Minnows appear scaleless, but in fact are evenly covered with fine scales and smooth skin over a streamlined body that is rounded in cross section. The body colour is brown on the back and silver or white beneath, with a prominent black line, formed from a series of overlapping 'dots', running along the midline for the length of the body. The fins are small, rounded and colourless. The mouth is also small and lacks teeth and barbels. Minnows spawn in shoals on the marginal gravels of rivers and large lakes multiple times throughout the summer. During spawning, male and female Minnows often form discrete shoals that come together to spawn. Males become spectacularly gaudy, developing white patches at their fin bases and with a kaleidoscope of emerald, red and gold colours across the body, females retaining their overall silver-brown colouration. The eggs and fry receive no parental care. Juvenile minnows feed extensively on

small invertebrates and algae. Growing and adult minnows feed opportunistically on a wide range of small food items, including invertebrates and plant matter.

The tench (*Tinca tinca*), also known as the 'doctor fish' (a nickname discussed in Chapter 3), is a fish of still waters, sluggish rivers and drains in the lowlands of the British Isles. Tench have a characteristically olive-green or occasionally brown colouration, fading to yellow beneath, over a rounded body shape. The flanks are smooth to the touch due to a dense covering of small scales overlain with a thick, gelatinous mucus. The fins are rounded and the eye small and red. The small mouth, which can be protruded to aid feeding, lacks teeth but has a small, single barbel at each corner. Tench are adapted to life in weeded still or sluggish waters, grubbing in soft sediments for plant and invertebrate food items. Tench spawn communally over dense vegetation in rapidly warming shallows during late spring or early summer when the water reaches 18°C. Groups of male tench, distinguished from females by their spoon-shaped ventral fins, chase and jostle gravid females. The female tench release many small, green sticky eggs, which adhere to water plants but that thereafter receive no parental care. Warm, shallow water is essential for the growth of juvenile tench, which feed extensively on a range of invertebrate and algal matter. Young tench remain elusive, living in dense vegetation as they grow, adult fish graduating to an omnivorous diet. A 'golden' strain of tench is bred for ornamental purposes.

The bitterling (*Rhodeus amarus*), also known as the bitterling carp, is a small fish (up to 16 g) introduced to British waters through releases from aquaria. They have become locally naturalised, with strongholds in still and sluggish waters in south Lancashire, Cheshire, parts of Shropshire and some of the Great Ouse catchment, though nowhere are bitterling abundant. Bitterling have deep, laterally compressed bodies with a short lateral line that peters out five or six scales behind the gill covers. The silver flanks contrast with the grey-green back, and a

**The tench.**

distinct metallic stripe extends from the middle of the flank to the base of the tail. The mouth is small and lacks barbels, pointing forwards or slightly to the underside of the blunt snout. The bitterling's favoured habitat is densely weeded regions of still waters and slow-flowing river margins with sandy or muddy bottoms where freshwater mussels occur, mussels playing a key

**The bitterling.**

role in the bitterling's unique life cycle, as will be discussed in Section 7.6. Bitterling have an omnivorous diet.

Rounding off the complement of fishes of the minnow and carp family in British waters are two species that are introduced and have become problematically invasive. These interlopers are the topmouth gudgeon (*Pseudorasbora parva*) and the sunbleak (*Leucaspius delineatus*), both occurring in pockets throughout England where they are cause for concern in terms of their impacts on native fish stocks. The topmouth gudgeon, also known in the aquatic trade as stone moroko or clicker barb, is a small fish (up to 16 g) native to flowing and still fresh waters from Japan to the Amur basin, where it feeds on invertebrates, fish and fish eggs. The topmouth gudgeon has an elongated, spindle-like body covered in prominent scales. The fins are not significantly elongated at the base. The snout is slender, the mouth oriented upwards and a prominent longitudinal pigmented line extends along the flank. The sunbleak, also known as the belica or motherless minnow, grows a little larger (up to 17 g) and is a shoaling surface-feeding fish from areas of Asia and eastern Europe, favouring slow-flowing and still waters. Sunbleak are bright silver in colour, easily identified by the incomplete lateral line that peters out shortly before the end of the pectoral fin. These two invasive species share a rapid growth rate, capacity to breed after only their first year and ability to spawn multiple times each year on leaves or stones throughout the spring and summer. They also exhibit parental protection of eggs by male fish and a tendency to eat the spawn and fry of other fishes, enabling them both to rapidly colonise new waters and to outcompete and prevent the reproduction of native species.

For those with particular interest, Appendix 4 tabulates key external identification features helpful in definitively distinguishing British freshwater cyprinid fishes.

**The topmouth gudgeon, also known as the stone moroko or clicker barb.**

**The sunbleak, also known as the belica or motherless minnow.**

## 5.2 FISHES OF THE SALMON FAMILY

The fishes of the salmon, trout, charr (alternatively spelled 'char'), freshwater whitefish and grayling family (Salmonidae, generally commonly referred to as the salmonids) are also widespread and not uncommonly encountered in the British Isles, though some species are rare and/or rarely encountered. Ten species occur in Britain, seven of them native (of which one is declared extinct). Two are introduced, and one has found its way to our shores from introductions elsewhere. These salmonid fishes are slender and streamlined with pelvic fins set far back on the underside and also a fleshy adipose (or fatty) fin towards the rear of the back. The mouth contains a single row of sharp teeth (with the exception of the grayling).

The most charismatic, if not the most widely distributed, of the British salmonids is the Atlantic salmon (*Salmo salar*). The Atlantic salmon is sometimes referred to as the 'king of the fishes' and is found in cleaner rivers particularly to the north and west. Atlantic salmon can grow large (up to 29 kg), with familiar, elegant silvery, streamlined bodies, large eyes and a mouth adapted to life as a fast-swimming predator. Mature male salmon develop a kype (or hooked lower jaw) as they run rivers from the sea to spawn, using it as a weapon to defend spawning sites. Atlantic salmon famously live out their adult lives at sea, returning after between one and four years to their natal rivers to spawn. (The breeding habitats of British freshwater fishes are considered in detail in Chapter 7.)

The adipose fin of a brown trout, a fatty fin situated on the back immediately in front of the tail fin, is characteristic of members of the salmon family (as well as some other fish families not found in British fresh waters). (Image © Dr Mark Everard.)

At the time of writing, another salmon species seems about to become established in British waters. The pink salmon (*Oncorhynchus gorbuscha*), also known as the humpback salmon, is native to the Pacific (northwestern US, western Canada, eastern Asia and the Far East of Russia), where it is the smallest (up to 7 kg) and most abundant Pacific salmon species. In August 2017, video footage emerged of pink salmon spawning in the River Ness, Scotland. Sporadic, largely unverified reports of pink salmon had come from a number of northern British rivers over preceding years, the first from Scotland in 1962. However, a number of pink salmon were caught and verified from the Tyne and other northeastern English rivers throughout 2017, as well as from rivers in the northern part of Ireland down as far as Galway. How did they get there? Russia had previously been stocking the species in rivers draining to the Barents Sea, and the fish had already started spreading to Norway and Iceland, so this spread to the British Isles is a logical progression of a species known to have a high fecundity. The oceanic form of the pink salmon is bright silver in colour but, as they run up streams to spawn, their colouration changes to pale grey on the back with a yellowish-white belly, male fish developing a characteristically humped back. At this time, it is not certain, though considered highly likely, that pink salmon will become established in British waters. The consequences of their naturalisation for competition with the endangered Atlantic salmon and other species are currently unknown.

Brown trout (*Salmo trutta*), also known by various local names including fario, bull trout, ferox or brownie, are fish particularly of cleaner and clearer rivers and large lakes throughout the British Isles. They also have a sea-going form, the sea trout, with a similar life cycle to the Atlantic salmon. Sea trout are known regionally as sewen in Wales, peel (or peal) in southwest England as well as in Ireland, finnock in Scotland and salmon trout in culinary circles. In ideal conditions, brown trout can reach 10 kg and sea trout can exceptionally reach 30 kg. The

**The Atlantic salmon.**

streamlined body of the brown trout is salmon-like, the head is small and the large mouth is armed with short, strong teeth. Body colouration varies with habitat and life cycle. The flanks of many freshwater trout are buttery-brownish with spots of varying colours, whilst sea trout take on a silver sheen. Sea trout have a larger mouth than Atlantic salmon, extending behind the eye and tend to have spots below the lateral line, also lacking the pronounced 'wrist' before the tail fin that is found in Atlantic salmon. Unlike rainbow trout, brown trout have between three and five spines on the leading edge of the dorsal and anal fins and lack a pink or red stripe along the flanks and spots on the tail characteristic of rainbow trout. Brown trout are in fact so adaptable that they were originally classified into many different species (a characteristic that will be picked up later in this book).

Rainbow trout (*Oncorhynchus mykiss*), also known as blue trout (a form produced in aquaculture) and with a sea-going form known as a steelhead salmon, are a fish originally from rivers flowing into the Pacific along the west coast of North America from Alaska to Mexico. However, they have been widely released from aquaculture and stocked for angling purposes across the world, including into British rivers and still waters. In a few British rivers, they have become naturalised, forming self-sustaining populations. However, most British rainbow trout still originate from stocking or releases from trout farms. Rainbow trout have a streamlined body covered in small scales with fine spotting and a characteristic wide pink or red stripe running along the flank from head to tail (though the blue trout form lacks this band). The jaw is lined with fine, strong teeth. In technical terms, rainbow trout are actually a form of charr rather than a true trout, lacking spines on the leading edge of both the dorsal and anal fins. Rainbow trout are predators, feeding on aquatic and terrestrial invertebrates and small fishes. Rainbow trout generally undertake short spawning migrations into suitable streams with a clean gravel bottom, male fish defending territories and female fish cutting redds into which they deposit eggs that are not then

**The brown trout.**

guarded. The life cycle is typical of other members of the salmon family. The sea-going form, the steelhead salmon, is infrequent in Britain and probably related to a few farmed rainbow trout migrating to sea rather than representing an established steelhead salmon population.

Brook trout (*Salvelinus fontinalis*), or American brook trout, are another American species that has been introduced a number of times to the British Isles, but naturalised in only a few scattered high-altitude lochs on the west coast of Scotland and in north Wales. Brook trout have the streamlined body and jaw lined with fine, strong teeth typical of members of the salmon family. The dorsal and anal fins lack spines at their leading edge, and the flanks are characterised by a combination of dark green marbling on the back and dorsal fin and by red

**The sea trout.**

**The rainbow trout.**

spots with blue halos on the flanks. Brook trout undertake short spawning migrations into suitable reaches of stream or lake margins with a clean gravel bottom, male fish defending territories and female fish cutting redds into which they deposit eggs that are not then guarded. The life cycle is typical of other members of the salmon family. A larger, sea-running strain is found on the east coast of Canada and the US, though no such migratory populations are known in the British Isles where specimens are generally very small but voracious in exploiting the sparse fare of invertebrate and fish fry found in the upland lochs where they occur.

At this point, it is worth mentioning another trout that has been introduced but never naturalised in the British Isles. The tiger trout is not a true species, but is a sterile hybrid between the brown trout and the brook trout. It is not known in the wild, either in Britain or the US where both parent trout species occur, but is produced in hatcheries for sport fishing. Tiger trout have been reported to grow faster than natural species and, though widely stocked for sport fishing in America, are no longer introduced to British waters. They are noted here not as a fish that the British naturalist may encounter, but purely as a curio.

Grayling (*Thymallus thymallus*), sometimes alluded to as 'the lady of the stream' for their graceful appearance, occur throughout mainland Britain in strongly flowing, clean rivers. Though not native to Scotland, grayling were introduced to the Clyde from Derbyshire in 1855 and have subsequently been introduced to many other Scottish rivers where they have become established and spread. Grayling are distinctive in appearance, with an elongated, spindle-like and streamlined body. Their flanks are silvery and iridescent, dappled with irregular dark spots interspersed with hints of purple, green and copper. The 'sail-like' shape of the large dorsal fin, particularly enlarged and coloured in male fish, is characteristic. Unlike other British salmonids, grayling have soft, toothless jaws, equipping them for feeding on invertebrates foraged from

The brook trout in its parr stage.

The tiger trout.

the stream bed or surface. Grayling are a shoaling fish, occurring in clean rivers. Curiously, they have an odour often likened to fresh thyme (from which their scientific name derives). In his 1653 book *The Compleat Angler*, Izaak Walton records, '*And some think that he feeds on water thyme, and smells of it at his first taking out of the water...*' though the reality is that the grayling feeds almost exclusively on invertebrates and, opportunistically, small fishes. Unlike other members of the salmon family, grayling spawn in the spring. However, like other salmonids, males defend territories on well-flushed gravel beds and females deposit unguarded eggs that are not subsequently guarded. Grayling grow rapidly but are not long-lived.

Arctic charr (*Salvelinus alpinus*), also referred to as char or in Wales as torgoch, are fishes of deep, well-oxygenated glacial lakes to the north and west of the British Isles. Arctic charr share

The grayling, female in the
background and male in the
foreground displaying its brightly
coloured, sail-like dorsal fin.

many of the characteristics of the salmon family, with an elongated, streamlined body and a small head and large mouth adapted for a mobile, predatory lifestyle. Unlike brown trout, Arctic charr have teeth only in the central forward part of the mouth. The body colour of Arctic charr varies dramatically with habitat, different strains once considered separate species. Arctic charr occur across northern North America, Europe and Asia. A cold-adapted species, they occur only in deeper waters of large glacial lakes of the northern and western British Isles and are considered as glacial relics. (In simple terms, this means that as ice sheets receded at the end of the last Ice Age, a few populations of these fish were stranded in deep, cool and well-oxygenated lakes.) Arctic charr are consequently extremely sensitive to organic and nutrient pollution, which reduces oxygen levels in the deep strata of the lakes in which they occur, and are consequently of significant conservation concern across their remaining British range. Arctic charr enter tributary rivers or migrate towards well-flushed lake margins during autumn or winter, when water temperatures reach around 4°C, spawning in well-oxygenated gravel beds. Schooling behaviour is abandoned at these times, mature males becoming territorial in their defence of ideal spawning gravels over which they coax gravid female fish to cut their 'redds' and deposit their eggs.

Rounding off the list of British salmonids are a bevy of whitefishes. These fishes have a scattered distribution in deep, clean lakes and are also considered glacial relics. There is consequently a lot of genetic diversity amongst populations and some disagreement about the distinctiveness of populations as sub-species. However, a general view is that there are two extant species: European whitefish (*Coregonus lavaretus*) and vendace (*Coregonus albula*). A further, estuarine and migratory species, the houting (*Coregonus oxyrinchus*), with a characteristically pointed snout, was declared extinct in 1982 after an absence of records for many years. Local names for European Whitefish vary by region, including powan (in Loch

Lomond), schelly or skelly (in Haweswater, Ullswater and Red Tarn in the English Lake District), gwyniad (in Llyn Tegid (Lake Bala) in Wales) and pollan (in Northern Ireland). The vendace, also known as the European cisco, has a more restricted distribution. Until recently, vendace were found in only two lakes in England, Bassenthwaite and Derwentwater in the English Lake District. It was concluded in 2008 that Vendace had become extinct in Bassenthwaite (a tale to be picked up in Section 11.3.2). Owing to a decline in the remaining Derwentwater population, the English government's nature conservation agency, Natural England, instigated a conservation project in the winter of 2004/2005 to translocate vendace from Derwentwater to refuge locations at Daer Reservoir (Lanarkshire) and Burnmoor Tarn (Cumbria) to establish safeguard populations. Two Scottish vendace populations, Mill Loch and Castle Loch, are now also considered extinct, but vendace stocked in the mid-1990s into Loch Skeen have become established as a conservation measure. These whitefishes have a similar, laterally compressed, spindle-like body with a bluish or dark green back and a conical head with a small ventral mouth. Whitefishes breed during the winter on marginal gravels around lake shorelines when the water is about 6°C. Fry and adults feed largely on invertebrates, sometimes rising up in the water column to feed by night. Populations are vulnerable to pollution, particularly by organic matter that suppresses oxygen levels and also to predation of eggs and fry when small predatory species such as ruffe are introduced.

## 5.3 THE PIKE

Although the pike (*Esox lucius*) is the only British member of the pike family (Esocidae), this native fish is widespread in rivers and still waters, as well as into estuaries and is therefore likely to be encountered by fish watchers. Pike enjoy something of a reputation as a large (up to 27 kg) predator sometimes known as the 'freshwater shark', also referred to as luce or *Esox*. Pike have a long body with numerous small scales along flanks that are usually mottled green

The European whitefish.

The vendace.

to provide camouflage. The pike's lifestyle as an ambush predator is further aided by its large eyes and cavernous mouth armed with sharp teeth. The anal and dorsal fins set to the rear of their streamlined, muscular flanks enable pike to accelerate explosively to intercept prey. The rounded fins are supported by soft rays. They ambush other fish (though they also seek out dead fish), as well as amphibians and small aquatic mammals and birds, and have pronounced cannibalistic tendencies. Pike spawn early in the year, from late February to May depending on conditions. Large female pike move into vegetated backwaters or shallows, generally accompanied by several smaller male pike. It is not uncommon for female pike to devour smaller male fish, assembling around them before spawning. Sticky eggs adhere to vegetation and receive no parental care. Juvenile pike are predatory from the fry stage, exploiting the fry of later-spawning fish species as well as their siblings.

## 5.4 THE PERCH FAMILY

Also charismatic and sometimes conspicuous in both running and still British fresh waters are the two native members of the perch family, with a third introduced species less often seen because of its propensity for nocturnal feeding and deeper waters. Members of the perch family possess two dorsal fins, the anterior of which has strong spines, with the posterior supported by soft rays. These dorsal fins are either separated or contiguous.

The perch (*Perca fluviatilis*), otherwise known as the European perch to distinguish it from the many species of perch across the Northern Hemisphere, is found in both still and flowing fresh waters and upper estuaries across the British Isles. Perch are a medium-sized fish (growing to 2.7 kg) with characteristic bold, black stripes on greenish, rough-scaled flanks. Perch have two dorsal fins, the front strongly spined and the rear, separated from the front fin, with soft rays. The large mouth lacks barbels and the jaws lack teeth, though the roof of the mouth is covered in small, backward-pointing vomerine teeth to retain engulfed prey. Perch commonly shoal and are often found near sunken branches and other vegetation from where they launch predatory raids. Shoals of hunting perch often betray their presence at dusk as shoals of small prey fish 'explode' at the surface to evade these raids. Perch spawn communally on hard submerged surfaces in the spring, the eggs and fry receiving no parental care. Juveniles feed predominantly on small invertebrates, progressing to a diet of larger invertebrates, amphibians and fish as they grow. Stunted populations of perch can occur in small still waters, where they breed prolifically but outstrip food resources.

Ruffe (*Gymnocephalus cernua*), also known as tommy ruffe or pope, are a small (up to 170 g) shoaling fish of slower running and still waters in lowland southern, eastern and central England. However, they have been spread more widely, where they can be a nuisance predating on the eggs and fry of other fishes. Ruffe have a generally silvery, mottled body

**The perch.**

colour with two dorsal fins, the front one spiny and conjoined with the rear fin which has only soft rays. The scales are small and the fins translucent. The large mouth lacks barbels, and the jaws lack teeth. Ruffe may be confused with small perch, but lack strong vertical stripes on their flanks. Perch also have a clear gap between the two dorsal fins. Ruffe spawn in shoals in shallow water during the spring. Females deposit their eggs, which receive no parental care, in sticky strands that adhere to vegetation and stones. Juvenile ruffe eat small invertebrates, with progressively larger animal prey taken as the fish grow. Introductions of ruffe into some waters where they are not native have resulted in serious threats to species of conservation concern, such as gwyniad and vendace, due to the ruffe predating their eggs and fry.

Zander (*Sander lucioperca*), or pike-perch, are native to continental Europe, from where they were introduced to Britain. Introduced populations were initially locally contained, but have now spread throughout the Ouse, Thames and Severn river systems, connected canals and adjacent reservoirs and are still spreading. The common name 'pike-perch' and some frequent assumptions suggest that these fish are a hybrid between the pike and the perch, sharing as they do intermediate characters of the two species. However, this is not the case: zander are a discrete species within the perch family. Zander grow relatively large (up to 9 kg), living in deeper reaches of rivers, in still waters and in upper estuaries. Zander possess two dorsal fins, the front supported by spines and the rear by soft rays with a distinct gap between these fins. The flanks are generally silver with a greenish tinge, covered in small scales. The eyes are distinctively large and reflective due to the presence of a *tapetum lucidum*, or a reflective layer behind the retina aiding hunting in low light. The jaws are armed with teeth but lack barbels, and the inner surface of the mouth is roughened by vomerine teeth. Zander are a largely crepuscular (active at dawn and dusk) or nocturnal predator, feeding primarily on other fish. As

The ruffe.

The zander.

an introduced predator, zander have the potential to perturb British freshwater ecosystems, so further introductions are discouraged. Zander spawn in late spring or early summer when water temperatures exceed 12°C, laying sticky eggs on hard substrates that may include rocks, plant stems and underwater tree roots. Male fish clean the spawning substrate and guard the eggs and newly hatched fry. Juvenile zander feed on zooplankton and other invertebrates, progressing to a mainly piscivorous diet.

## 5.5 OTHER SMALLER FISHES

Minnows and gudgeon, bitterling and ruffe, are interesting elements of Britain's smaller fish fauna and have already been considered within the families to which they belong. In this section, we will turn to the rest of the smaller species one might encounter in British fresh waters.

Prime amongst these are the two native species of freshwater sticklebacks (a third British species is entirely marine), belonging to the stickleback and tubesnout family (Gasterosteidae). The bodies of these small fishes (up to 20 g) are generally elongated and stiffly held and lack scales or are armed with scutes (a few large bony scales) along the sides. The mouth lacks barbels and is generally small, at the end of a narrow tapering snout. There are a number of well-developed dorsal spines in front of the dorsal fin. The three-spined stickleback (*Gasterosteus aculeatus*), taking its name from the front dorsal fin being reduced to three stout spines, is commonly encountered in still waters and slower-moving margins of rivers as well as into estuaries and even coastal waters. During the spring and summer breeding season, male three-spined sticklebacks develop emerald flanks and a red breast, while females remain a browny-silver colour. Three-spined sticklebacks are hardy, tolerant of significant pollution and can form dense shoals. In some rivers, particularly spate rivers, three-spined sticklebacks overwinter in estuaries or coastal waters. The ten-spined stickleback (*Pungitius pungitius*), also known as the nine-spined stickleback, has the front dorsal fin modified into a row of nine or ten shorter stout spines, occurring in river edges, small streams and still waters throughout Britain except northern Scotland and also Ireland. Ten-spined sticklebacks have a more regular dull or silvery body colour and are an occasionally shoaling fish favouring well-vegetated habitats in river edges, small streams and still waters. However, they are intolerant of brackish water. Both stickleback species, sometimes locally known as prickleback or stickybag, are carnivorous, feeding on small invertebrates and fish fry. They share a similar life history, male fishes building nests of vegetation glued together using secretions from their kidneys rich in the glue protein spigin. Males then attract females with a characteristic 'zig-zag' dance, driving them off after their eggs are laid in the nest. This process may be repeated with three or four different females. Male sticklebacks then become devoted fathers, fanning water through the nest, removing dead eggs and caring for the fry for a few days after they become free-swimming.

The bullhead (*Cottus gobio*) is the only British species of the sculpin family (Cottidae) and is characterised by a scaleless body, though prickles occur on the gill cover and elsewhere. They go by a wide variety of alternative and local names including miller's thumb, bullyhead, mullyhead and wayne. Bullheads are small (up to 20 g), bottom-dwelling fishes lacking a swim bladder. They live under stones or woody debris, not uncommonly living out their whole adult lives in the same territorial 'cave'. Bullheads are common in the gravel or rocky bed of rivers and wave-lapped edges of larger still waters throughout mainland Britain, but are absent from Ireland. The bullhead is so named for its large and broad head, with a big mouth and a pair of small eyes positioned on the top. The mottled brown body tapers away behind with two dorsal fins, the front dorsal fin shorter with spines and the rear one soft, and a long anal fin beneath. A large pair of mottled pectoral fins match body colouration. Bullheads are exclusively carnivorous, feeding on invertebrates and fish fry. They spawn during the spring, male fish

The three-spined stickleback.

The ten-spined stickleback, also known as the nine-spined stickleback.

enticing neighbouring females into their territorial caves to deposit around 100 sticky eggs in clusters generally on the cave's ceiling. After fertilising the eggs, the male drives off the female and then nurtures the eggs through to hatching, caring for the fry until they venture out to locate and establish their own territories.

Found in the same habitat, and sometimes sharing the same stone refuge as the bullhead, is the stone loach (*Barbatula barbatula*), also known as the loach, stoney, beardie or groundling. Stone loach are the only British species of the stone loach family (Nemaceilidae), possessing elongated, scaleless bodies with small, inferior mouths surrounded by at least three pairs of elongated barbels. Stone loach are small fish (up to 20 g) found throughout the British mainland except Scotland. The body colour of the stone loach varies with habitat, but is usually dull yellow-brown with irregular blotches. The fins are rounded, the dorsal fin set to the rear of the body. Six barbels surround the mouth, two pairs beneath the tip of the snout and one pair at the corners of the mouth. Stone loach may be confused with spined loach, but have longer barbels and lack backward-pointing spines beneath the eyes. The stone loach inhabits running waters including small streams and, occasionally, the shorelines of lakes. They are secretive by day, concealing themselves under stones and dead wood on sandy, muddy or stony bottoms, emerging to feed primarily in the dark on a range of small invertebrates. Adult fish are largely solitary, though favourable refuges may harbour other loaches as well as bullheads. Stone loach spawn in the spring, females shedding clusters of sticky eggs amongst gravel, submerged stones and plants. Females have sometimes been reported guarding the eggs. Habitat in which to hide is important for the growth and survival of this secretive fish.

The spined loach (*Cobitis taenia*), also known as the loach, spiney or groundling, belongs to the loach family (Cobitidae) and is characterised by a spindle- or worm-like body with a small, inferior mouth surrounded by at least three pairs of barbels. Spined loach are small (growing up to 30 g), elongated and strongly laterally compressed, with minute scales and light brown colouration with nineteen brown spots along the flanks. Three pairs of short barbels surround the small mouth, and a strong, retractable double-pointed spine is located

**The bullhead. (Image © Dr Mark Everard.)**

**The stone loach.**

in a skin pouch below and in front of each eye, a key feature distinguishing them from stone loach. The spined loach is one of Britain's smallest and rarest native fishes, almost entirely restricted to a few catchments in eastern England. Spined loach are not highly mobile, living in dense submerged vegetation where they remain by day, emerging by night to feed on small bottom-living invertebrates as well as some vegetable matter. Spined Loach spawn in spring, depositing unguarded eggs on submerged plants, roots or stones. The fry are very small, and the generally small size of adults too necessitates dense vegetation for survival and reproduction.

**The spined loach.**

## 5.6 THE ELUSIVE EUROPEAN EEL

The European eel (*Anguilla anguilla*), more generally referred to simply as the eel or the snig (mainly smaller specimens), is the sole British freshwater member of the freshwater eel family (Anguillidae). Eels are adaptable, living in a wide variety of still and running, estuarine and fully saline waters. Their snake-like bodies appear scaleless but are covered in minute scales coated in a thick layer of slime making the fish smooth to the touch and difficult to hold. The dorsal fin runs most of the length of the body, contiguous with the caudal fin and the anal fin. Ventral (or pelvic) fins are absent, but the pectoral fins are well developed. European eels have a large mouth, which lacks barbels. European eels often evade detection, burying in soft sediment or inhabiting caves or dense vegetation, becoming actively predatory by night. The eel has a remarkable life cycle. Eggs assumed to be laid in the Sargasso Sea (more on this in Section 10.2) produce flattened larvae known as *leptocephali* that drift for two years on the Gulf Stream, arriving on European shores whereupon they metamorphose into transparent 'glass eels', which subsequently darken into elvers. Some remain in coastal waters and estuaries, whilst others ascend rivers and other water bodies including streams, lakes, reservoirs, ponds and ditches, even travelling over land on wet nights. Late in the year, often on dark nights with a new moon and after as many as twenty years, maturing eels, which have turned silvery, run back down rivers to sea to complete the cycle. Vulnerable across so many stages in its remarkable life cycle, which we will return to later in this book, populations of European Eel have dramatically declined in recent decades, and the species is now of major conservation concern.

**The European eel.**

## 5.7 BRITAIN'S ANCIENT LAMPREYS

Three species of the lamprey family (Petromyzontidae) are found in British fresh waters. All of them breed in fresh waters and have a freshwater larval stage. However, two of the species take to sea during their adult phase. The bodies of all lampreys are eel-like, but lack both scales, paired fins and jaws, and with a cartilaginous skeleton. Lampreys also have only a single nostril on the top of the head. For this reason, the lampreys are a good example of a fish that is not, in biological terms, a true fish. They belong to a group of fish known as the Agnatha (the Greek for 'no jaws') which, as we know from the fossil record, are the oldest living fishlike animal. Lacking jaws, the mouth of adult lamprey instead comprises simply a circular disk. Larval forms, known as ammocoetes, are characterised by a line of circular gill openings behind the eye, living buried for several years in river silt where they filter-feed on fine suspended matter in the passing water or graze on detritus before metamorphosing into eel-shaped adults that are initially shorter than their larvae.

The brook lamprey (*Lampetra planeri*) is also known as Planer's lamprey, the European brook lamprey, nine eyes, pride, sandpride or brookie. These small fishes (up to 60 g) occur in rivers throughout the British Isles except the far north of Scotland. Brook lamprey ammocoetes are rarely longer than 15 cm (6 inches), are toothless and blind with little pigmentation and possess two dorsal fins that merge into the tail fin. After living several years as ammocoetes, brook lampreys metamorphose into eel-shaped adults that are shorter than their larvae. Adult brook lampreys are slightly metallic in colour, with a pair of functional eyes and a small sucking disc with blunt, weak teeth. However, unlike the other two British lamprey species, brook lampreys do not migrate and nor do they feed as adults, instead spawning communally in river gravels shortly after metamorphosis, and dying shortly thereafter. The seven small, round gill openings on either side of the head were once thought to be additional eyes (hence the alternative name 'nine eyes').

**The ammocoete larva of a lamprey.**

The river lamprey (*Lampetra fluviatalis*) is also known as the lampern, juneba, stone grig or lamper eel. These are larger fish (up to 150 g) found as ammocoete larvae in the silt of river margins and as adults in rivers and estuaries throughout the British Isles. After metamorphosis, adult river lampreys develop a round, disc-shaped mouth lacking jaws at the front of a drab, slate-grey or brownish eel-like body lacking scales and paired fins. Behind the pair of functional eyes are seven small round gill openings either side of the head. The river lamprey has a similar life history to the brook lamprey to the point of metamorphosis from the ammocoete, after which the river lamprey grows on and migrates down river to live out its adult life in the estuaries of large rivers. Here, it subsists as a parasite, attaching itself to the sides of other fishes and feeding on flesh and body fluids rasped off with circular rows of teeth in the mouth disk. Parasitised host fish normally die. Mature river lampreys run rivers to spawn in late spring or summer, adults working in pairs or groups to move stones to create spawning depressions into which sticky eggs are laid. There is no parental care, and adult river lampreys die shortly after spawning.

The sea lamprey (*Petromyzon marinus*), also known as the marine lamprey or lamprey eel, is a larger fish (up to 2.5 kg) found throughout the British Isles but mainly in larger rivers, estuaries and coastal waters of England and Wales. Sea lamprey ammocoetes have extensive black pigmentation. They metamorphose into robust, eel-like adults with a cartilaginous skeleton, a round, disc-shaped mouth lacking jaws, a large pair of functional eyes and lack scales and paired fins. Seven small round gill openings are found on either side of the head. The sea lamprey has a similar life history to the river lamprey, migrating to sea after metamorphosis to live as a parasite, attaching to and rasping off flesh and body fluids from prey fishes with circular rows of teeth in its jawless mouth disk. Mature sea lampreys run rivers to spawn in late spring or summer, adults working in pairs or groups to move stones to create spawning depressions into which sticky eggs are laid. There is no parental care, adults dying shortly after spawning.

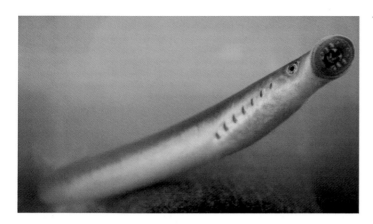

## 5.8 THE BRITISH SHADS

There are also two British species of shad, in the herring, shad, sardine and menhaden family (Clupeidae), both of which have an adult marine stage but which run freshwater rivers to breed. The body of both these shad species is strongly laterally compressed and generally herring-like, covered by large, round and smooth scales. The head is scaleless, the teeth are small or minute, and the fish also have long gill rakers (extensions of the gills inside the mouth to aid filtering fine food from the water). Both shads are highly localised and increasingly uncommon in the British Isles, running a few fast and clean western rivers and are sometimes found in estuaries. Both feed on small fishes and crustaceans of a size that matches the fish as they grow. These two shad species migrate up rivers in the spring to spawn in large shoals by night in fresh running water over sand or gravel bottoms, the allis shad reported to do so at higher temperatures (22°C –24°C) than the twaite shad (10°C–12°C), though this is contradicted by conditions in the British rivers in which they spawn. Spawned fish return to sea, fry progressively moving downriver to reach estuaries during their first summer.

The twaite shad (*Alosa fallax*) grows up to 1.5 kg, has a dark spot behind the gill with seven to eight spots along the flank and the gill rakers are shorter than the gill filaments (the outer, blood-rich filaments of the gill used for breathing). The allis shad (*Alosa alosa*) is superficially very similar, growing up to 4 kg also with a dark spot behind the gill but none along the flank, and the gill rakers are nearly or as long as the gill filaments.

THE COMPLEX LIVES OF BRITISH FRESHWATER FISHES

**The twaite shad.**

**The allis shad.**

## 5.9 THE NOT-SO-COMMON STURGEON

Despite its name, the common sturgeon (*Acipenser sturio*), or European sturgeon, is extremely rare and is now listed as Critically Endangered on the International Union for Conservation of Nature (IUCN) Red List (a classification scheme of the risk of extinction of species of plants and animals). It is the only species in the sturgeon family (Acipenseridae) found in British waters (with the exception of some smaller sturgeon species found occasionally as ornamental species). The body of the common sturgeon is elongated and armed on the sides by five rows of scutes (large armoured scales). Common sturgeon can grow very large (up to 400 kg). The colouration is olive-blue on the back, fading to paler on the underside. The mouth of the sturgeon is small, toothless and protrusive, located on the underside of an elongated snout, with four small barbels in front. The common sturgeon is anadromous, living its adult life (about which little is known) at sea and returning to rivers to breed often after many years. For this reason, the common sturgeon is occasionally encountered in estuaries and lower, large rivers, though it is a scarce and generally solitary visitor to British waters. Female common sturgeon deposit 800,000–2,400,000 small dark, sticky eggs at each spawning on sand, gravel and stones. Juveniles return to estuaries and the open sea, growing slowly and maturing after seven to nine years. Some specimens can live as long as 100 years. The common sturgeon was once plentiful; until comparatively recent times, it was reported as being found in a great number of British rivers including the Thames. (An image of an angler appears to involve a sturgeon in a mural on the wall of St Etheldreda's Church in Horley, Oxfordshire, dating back to at least the middle seventeenth century.) However, early angling literature makes only sparse mention of sturgeon, with Walton's 1653 *The Compleat Angler* entirely omitting mention of this fish.

**The common sturgeon.**

## 5.10  THE CURIOUS CASE OF THE EXTINCT BURBOT

Another freshwater fish species that is rare to the point of being declared extinct from its former range in eastern England is the burbot (*Lota lota*), also known as the mariah, 'the lawyer' or eelpout. The burbot is the only freshwater member of the hake and burbot family (Lotidae), a group of cod-like fishes. Reflecting their marine ancestry, burbot possess three dorsal fins, a single anal fin and a rounded tail fin, as well as a single chin barbel. Burbot are large fish (up to 34 kg) and are widespread from Canada and the northern US and across northern Europe and Russia, but seem to have disappeared from British waters between the 1950s or 1960s. In the 1970s, a reward was offered for any burbot captured from British waters, but this prize still remains unclaimed despite persistent rumours of unverified captures and relic populations.

Burbot are fishes of sluggish rivers and still waters. Burbot spawn in the winter, often under ice, with mature fish migrating into shallow margins to spawn communally at night on sand or gravel beds. The tiny eggs, one to three million released by each female, contain an oil globule enabling them to float off into the water column. Many die, but those that survive produce minute burbot fry that feed on small planktonic animals, progressing to larger invertebrate prey as they grow.

The burbot. (Image © Viktor Vrbovsky.)

## 5.11 CATFISHES, GREAT AND SMALL

Two species of catfish have been introduced into British waters, one a large species introduced deliberately and the other a small American species probably arriving as an aquarium release. Both are potentially problematic.

The wels catfish (*Siluris glanis*) is often simply referred to as catfish, moggie (particularly small catfish which are also referred to as kittens) or whiskers. The wels catfish, native to fresh and brackish waters of western Asia and continental Europe, can grow very large (up to 300 kg in ideal conditions) and is a voracious predator. This sole British member of the sheatfish catfish family (Siluridae) has been introduced and locally naturalised in some southern still waters of the British Isles, but can also inhabit sluggish rivers. The wels catfish has small, widely spaced eyes, and the body is elongated, lacking obvious scales, and tapers behind a large mouth like a giant tadpole. There are no barbels on the snout, but there are one or two pairs of elongated 'whiskers' on the lower jaw and also an elongated pair of maxillary barbels. Body and fin colour is dark, generally a mottled grey, but paler beneath. The dorsal fin is small, lacking spines, but the anal fin is very long. Wels catfish are generally inactive by day, hiding in cover, but emerge as voracious nocturnal predators that feed in warmer months on almost any fish, fowl or mammal prey. Wels catfish breed in early summer in vegetated margins of lakes and large rivers, laying eggs on mounds of leaf litter which are guarded by the male catfish. Juvenile wels catfish are superficially similar to tadpoles, but grow rapidly in warm water where they feed on a variety of invertebrates, progressing to a wholly carnivorous diet. Given their large size and voracious appetites, further introductions of wels catfish are unwise.

**The wels catfish.**

**The black bullhead. (Image credit: Rostislav Stefanek/Shutterstock.com.)**

The second of the introduced catfish found in some British waters, mainly in southern England, is the black bullhead (*Ameiurus melas*) from the North American freshwater catfish family (Ictaluridae). The black bullhead is no relation, nor is it even closely related, to the native bullhead (*Cottus gobio*). Black bullheads can grow to moderate size (up to 3.6 kg) and have become naturalised in pockets throughout England, presumably as a result of incautious aquarium releases. Black bullheads have black, scaleless bodies, possess eight barbels around the mouth, and are armed with three long, sharp spines, one each on the leading rays of the pectoral and dorsal fins. Black bullheads are widely introduced across Europe where they can form dense populations and can be problematic as they are voracious and omnivorous, potentially outcompeting other species and eating their spawn and juveniles.

### 5.12 THE RAINBOW PERCH

The pumpkinseed sunfish (*Lepomis gibbosus*), also known simply as the pumpkinseed, pond perch or sunny, is from the American sunfish family (Centrarchidae). Pumpkinseeds are smaller fish (growing up to 180 g) and have become established in a number of still waters and calm river margins across Britain, locally from Somerset to as far north as County Durham. They are common now where introduced, naturalised and where they have spread

across continental Europe, one of the French names being 'perche arc-en-ciel' (or rainbow perch). Pumpkinseed sunfish have a round body profile and are laterally compressed, their shape contributing to the common name 'pumpkinseed', as well as the name 'panfish' applied to this group of fishes in North America. These fish have electric-blue highlights across the flanks and strong blue, green and red colouration around the head, with a long dorsal fin consisting of a spined front fin fused into a soft-rayed rear fin. They have a predatory diet. They are thought to have been introduced into English waters by releases from the ornamental fish trade, but also potentially from research facilities where these hardy fish are commonly used for scientific experiments.

## 5.13 MARINE INTRUDERS

This then leaves us with a number of species entering the lower reaches of British rivers that are transient visitors from the sea, particularly during the summer months, rather than being true freshwater fishes. These 'marine visitor' species, considered briefly in this section, include three species of mullet, the European seabass, the flounder, the smelt and the sand smelt.

There are 72 species in the mullet family (Mugilidae) across the world. The mullets are predominantly marine and estuarine fishes possessing a short front dorsal fin supported by four stout spines and a rear dorsal fin supported by soft rays. The pectoral fins are situated high on the flanks with a lateral line, when present, that is barely visible. The mouth of each British species lacks teeth and is upward-oriented with rubbery lips at the front of large head that is flattened above as an adaptation for swimming and feeding near the surface. These schooling fishes have extremely long intestines to aid digestion of their diet of fine algae, including diatoms and detritus, grazed from bottom sediments with their rubbery lips. Three mullet species are found in British estuaries in the summer, some penetrating further up river systems. All three breed at sea during the late summer or winter, laying pelagic and non-adhesive eggs, the juveniles drifting inshore and often inhabiting estuaries. The mullets are largely inshore species in the summer months.

The largest species (up to 6.5 kg) is the thick-lipped grey mullet (*Chelon labrosus*), which enters brackish lagoons and fresh waters during the summer when it migrates northwards as waters warm. The upper lip is thick, its greatest depth equal to at least half the eye diameter, and several rows of small, dark papillae are present on its lower border. The thin-lipped grey mullet (*Liza ramada*) grows a little smaller (up to 3.2 kg) and is similar in many ways to the thick-lipped grey mullet, but with smaller and thinner lips (always less than half the diameter of the eye) lacking papillae (small projections). The golden grey mullet (*Liza aurata*) has all the generic features common to the mullet family, but is distinguished by a golden spot on the gill cover giving the species its common name, and also, the pectoral fins are longer than in the other British mullet species, reaching the eyes when folded forward. There is also no black spot at the base of the pectoral fin. The mouth is small, and the upper lip is thin (always less than half the eye diameter) and lacking papillae.

**Shoal of thick-lipped grey mullet grazing gravel surface in an estuary. (Image © Dr Mark Everard.)**

The golden grey mullet. (Image © Dr Mark Everard.)

The European seabass (*Dicentrarchus labrax*), usually simply referred to as the 'bass' in Britain where it is the only member of the temperate bass family (Moronidae), is also locally known as the sea dace for its bright silvery colouration and high activity level. The European seabass is a swiftly swimming predator of coastal waters and estuaries with a graceful, elongated body that is silvery-grey to bluish on the back, silver on the sides, and white on the belly, sometimes tinged with yellow. The front dorsal fin is armed with 8–10 stout spines, with the rear dorsal fin held aloft by a single front spine and 10–13 soft rays. There are also two spines on the rear of the gill cover (or operculum). The jaws lack teeth, but a band of small vomerine teeth is present on the front part of the roof of the mouth, serving to hold prey – mainly fish but also

larger invertebrates such as crabs – once captured. Bass can grow large (up to 19 kg). They are notably migratory in behaviour, moving northwards and inshore with warming weather in spring and retreating southwards again in the winter, frequently migrating into brackish and sometimes fully fresh water during summer excursions. (It is worth mentioning in passing that there is another bass species, the spotted seabass, *Dicentrarchus punctatus*, native to marine and brackish waters of the coastal eastern Atlantic Ocean, though this is a rare visitor to our shores and will not be considered any further.)

Flounder (*Platichthys flesus*) occur commonly around British coasts. They are the only British member of the 93 species of the righteye flounder family (Pleuronectidae) that commonly penetrates up rivers, predominantly during the summer months, but they are a common estuarine resident during the autumn and winter. As Izaak Walton put it nicely in *The Compleat Angler*, '*And, scholar, there is also a FLOUNDER, a sea-fish which will wander very far into fresh rivers, and there lose himself and dwell…*' Flounder spawn in marine waters in the spring, but routinely enter regions of British estuaries with fresh water outflows and may penetrate considerable distances upstream into fully fresh water during the summer months, feeding on benthic invertebrates and small fishes. Both eyes of the flounder are normally on the right side of the body, and the fins lack spines, including the dorsal fin that extends onto the head. The pigmented upper surface of the body is capable of remarkable colour changes to match the muddy and sandy bottom sediments of the river or sea bed. The lateral line is straight, but slightly rounded over the pectoral fins.

The smelt (*Osmerus eperlanus*), also known as the sparling or European smelt, is a small fish (up to 250 g), the only member of the smelt family (Osmeridae) recorded in British waters. Smelt are often residents in larger estuaries, entering the lower reaches of rivers. They have a

The flounder.

single dorsal fin lacking spines but supported by soft rays. The head and snout are pointed. The lower jaw has small teeth and reaches to the hind margin of eye also projecting a little beyond the tip of the upper toothed jaw. There is also an incomplete lateral line, and a pronounced silvery stripe runs along the bright flanks. Another characteristic feature of smelt is that living and freshly caught fish are said to smell of cucumber. Although the smelt is listed in the Reverend W. Houghton's 1879 *British Fresh-Water Fishes* as a species of fresh waters, and some freshwater smelt populations occur across the Europe, British smelt are inherently marine. They enter estuaries and occasionally lower rivers to spawn between February and April, adult fish then returning to sea with the young fish also moving down to estuaries after hatching.

The smelt.

**The sand smelt.**

The sand smelt (*Atherina presbyter*), also known as the little sand smelt, is a member of the silversides family (Atherinidae) and is not related to the smelt (*Osmerus eperlanus*) despite the similarity of common name. Sand smelt are a small (up to 100 g) shoaling species of inshore and coastal marine waters, favouring estuaries and penetrating the lower reaches of rivers. The body of the sand smelt has an intense silvery line, often outlined in black from head to tail along a silvery, elongated and laterally compressed body that is covered in relatively large scales. The body lacks a true lateral line. These fishes have two widely separated dorsal fins, the first with flexible spines and the second with one spine followed by soft rays, and a large pair of pectoral fins. The eyes are large, their diameter equal to the length of the snout, and the mouth is upturned. Sand smelt are gregarious, favouring low salinity water and feeding primarily on zooplankton, particularly small crustaceans and fish larvae. Isolated freshwater populations of the closely-related big-scale sand smelt (*Atherina boyeri*) are known in several Italian and Spanish lakes and in some lower river reaches around the Mediterranean, but this species does not occur in British waters. Sand smelt reproduce in spring and summer, favouring intertidal pools and coastal lagoons.

### 5.14 IMAGINARY FRESHWATER FISHES

Lastly, we turn to a subset of freshwater fish species that used to appear in some older works but that were, as we now perceive them, misclassified. These include various trout and charr species, the graining, the azurine and the dobule (the latter three formerly considered to be cyprinids).

The species we know and love today as the brown trout or the sea trout was formerly classified into multiple species, as were local varieties of the Arctic charr. We will return to this topic when considering the significance of regional strains of the same species in Section 11.1.

The graining, classified as *Leuciscus lancastriensis* in William Yarrell's text *The History of British Fishes*, first published in 1836, was based on a single specimen that Yarrell had examined from a small tributary of the Mersey near Formby. By any, let alone modern, standards, this is hardly the most robust scientific basis upon which to describe a new species. The 1843 book *The diary of A. J. Lane: With a Description of those Fishes to be Found in British Fresh Waters* provides a description of the graining:

> *In form somewhat like a Dace, but much rounder in the belly, tail deeply forked, top of the head and back a sort of drab coming down the sides and terminating rather suddenly; beneath this and the belly much lighter, eyes whitish, cheeks white, or yellowish, scales much smaller than in the Dace.*

> *This fish is believed to be confined to the waters of Lancashire. They are said to rise very freely to flys and also take baits as for Roach.*

The graining looked like a brownish-tinged dace, and in reality, that is exactly it seems to have been. The species subsequently disappeared from scientific and popular publications shortly after it featured in the Reverend W. Houghton's 1879 book *British Fresh-Water Fishes*, which includes an otherwise beautiful painting of the graining by the artist Alexander Francis Lydon. The British ichthyologist C. Tate Regan finally and definitively dismissed the separate identity of the graining in his 1911 book *The Freshwater Fishes of the British Isles*.

**The graining and dace, as painted by Alexander Francis Lydon.**

Also described in Houghton's book, and painted by Lydon, is the azurine. This fish had been scientifically defined as *Leuciscus caeruleus* Yarrell 1837, and formerly also known as *Cyprinus caeruleus* Yarrell 1833. The azurine was also known as the blue roach or lemon-finned rudd and was reputedly found in very limited localities within the town of Knowsley, Lancashire. *The Diary of A. J. Lane: With a Description of those Fishes to be found in British Fresh Waters*, published in 1843, describes the azurine thus:

> *In shape like a Roach but rather longer, tail long and forked; head, back and sides slate color passing into silvery white.*
>
> *Never exceed one lb in weight; is hardy and tenacious of life.*

The Reverend W. Houghton adds to this in *British Fresh-Water Fishes* (1879), after having examined many preserved specimens of this fish, that:

> *The Rudd, like fish generally, is subject to variety, and it would appear that the Azurine, or Blue Roach, first described by Yarrell under the name of* Leuciscus caeruleus *(Linn. Soc. Trans, vol. xvii. p. i. p. 8.), is merely a variety of the Rudd.*

AZURINE    DOBULE    RUDD

**The azurine, dobule and rudd, as painted by Alexander Francis Lydon.**

The azurine was also identified by William Yarrell on what can best be described as limited grounds, namely an anomalous fish with a bluish sheen. Other scientists agreed that this fish was a local variety of the rudd though, and perhaps unsurprisingly, specimens were no longer subsequently found in the Knowsley area. The azurine is now considered to be simply a synonym for the rudd (*Scardinius erythrophthalmus*).

The dobule, or dobule roach, was another fish that William Yarrell enthusiastically classified, again in a less than scientific manner from a single specimen reported to be taken from the brackish waters of the Thames estuary. In considering the dobule in *British Fresh-Water Fishes*, the Reverend W. Houghton appears to have had doubts about this fish too, as he writes:

> The Dobule Roach, a single specimen of which Yarrell took with the mouth of a White-bait net in the Thames below Woolwich, and which is regarded by Günther as a small Dace, is thus in its principal characteristics described by Yarrell...

The 'species' was once thought to exist elsewhere across Europe. Initially, enjoying the scientific name *Cyprinus dobula* Linnaeus 1758, the fish was later reclassified as *Leuciscus dobula* (Linnaeus, 1758) reflecting its dace-like affinities. The next move was to realise that the dace-like dobule was, in reality, a dace.

# FRESHWATER FISH
# ECOSYSTEMS

## 6

Other than in aquaria and garden ponds, Britain's freshwater fishes have to be considered as integral elements of the ecosystems of which they are a part. This includes not merely the mix of omnivorous, predatory and other fish species, but the plants, invertebrates, birds, mammals, fungi, amphibians and reptiles with which they mesh as part of complex food webs. It also includes the various life stages from egg to juvenile and adult of these fishes, some of them long-lived, all of which have different interactions with and influences upon their supporting environment. Fish are inextricably part of nature, mixed and healthy populations co-dependent with the functioning, vitality and resilience of habitats in which they occur.

It then makes more sense to think of our wealth of fishes in terms of the ecosystems of which they are a part, rather than analysing each in fine but isolated detail. This makes better sense of unique adaptations to their particular life habitats and niches within wider ecosystems. In this chapter, we therefore look at the places in which fish live and how they go about the often precarious task of living there.

### 6.1 OUR MOIST ISLAND NATION

The British people have a well-known obsession with the weather, a constant feature, or at least a punctuation, of many conversations. But then, we do have rather a lot of it.

To put this in global perspective, I recall a hazy April day in rural Rajasthan (India's northwestern desert state), asking the village headman if rain was on the way. After a moment's reflection, gazing skywards, he shook his head and said, '*No, not until September*' (and he was more or less right). Many parts of southern Europe, mid-continental US and regions of the tropics live with such seasonal certainties, including protracted dry periods when furniture is moved outside where it will not be spoiled by dew and rain in the low humidity until the weather breaks in its allotted season.

But the British Isles just happen to be situated where southwesterly prevailing winds, carrying warmth and moisture from the Gulf Stream, butt up against the polar air space. At this interface, moisture rising from warm oceanic currents to the south and west condense, leading to the frequent formation of clouds with associated high rainfall and constant background humidity. As the jet stream – a fast-moving ribbon of air flow running west to east some 9–16 km above the Earth's surface – kicks south, cool polar conditions prevail. As it kicks north, we enjoy the warmer, moist wash of Atlantic flows.

Where these two air cells meet, clouds are a given. We are, in geographic terms, blessed by a high frequency of overcast skies and a mixture of forms of precipitation from rain to sleet, snow, hail and drizzle. The weather we get on any particular day and place is the luck of the draw, dictated by the current status of this eternal aerial tussle.

A diverse cloudscape, a natural gift of the geographical location of the British Isles. (Image © Dr Mark Everard.)

Perhaps we don't appreciate, then, what a great place the British Isles are to watch clouds. We have so many and such a wide diversity of them. At high altitude, we enjoy the delicate filaments of cirrus, fibrous veils of cirrostratus and patchy sheets of cirrocumulus. At mid altitudes, we have sheet-like altostratus, laminated rolls of altocumulus and the thick, rain-bearing continuum of nimbostratus. Then, at lower elevation, we see the familiar, cauliflower-like mounds of cumulus; uniform layers of stratus; towering cumulonimbus with rising anvil heads generating hail, tornadoes and heavy rain and the honeycombed layer of stratocumulus. As with many of nature's gifts, perhaps we Brits take our rich cloudscape too much for granted?

Consequently, the British Isles are generally well watered, though the pattern of precipitation is far from uniform. Geographical location combines with the heterogeneity of geology and topography to generate mixed weather. As moist, oceanic air flows hit land and are deflected upwards, cooling condenses water into droplets. This is why, for example, we often see fog rolling in on the coast or hanging over hills. Meanwhile, England's southeast is in a 'rain shadow', where prevailing west and southwesterly air flows have already dropped their moisture and rainfall is lowest exactly where the human population is at its greatest density, with rainfall per capita famously lower than in Libya. So, on a cloudy day, or when the rain washes out our plans, perhaps we should reflect on our global geographical context and relative freedom from water poverty. The British cloudscape and propensity to dampness is, in reality, something special to be treasured rather than about which to complain!

A consequence of our geography is that these factors combine to create a wealth and diversity of rivers, pools, lakes, bogs, fens and other wetlands. And, consequent from our high population density and long history of industrialisation and intensive agriculture, there are multiple and often locally severe human pressures upon them.

## 6.2 RIVERS AND STREAMS

In the 1950s, there was an emerging view that European rivers comprise a sequence of zones. A popular river classification published by Marcel Huet in 1959 distinguished five major ecological zones, separated by longitudinal section (the slope of the stream bed) and the cross section of the stream and its valley, estimating their fisheries potential. Simply, this comprised:

- Fishless: Rapid headwaters that are too turbulent to support fish;
- Trout zone: Fast waters suiting the needs of trout, also with bullhead, minnow and stone loach;
- Grayling zone: Quick flows, slower than the trout zone, supporting faster-water species including grayling, dace, gudgeon and minnows;

The turbulent 'trout zone' of an upper river. (Image © Dr Mark Everard.)

**The sinuous track of a lowland river meandering within its floodplain. (Image © Dr Mark Everard.)**

- Barbel zone: Moderate currents meeting the needs of barbel, shared with fish such as chub, dace and bleak; and
- Bream zone: Sluggish flows, dominated by shoals of bream and other deeper-bodied fish, including smaller species such as roach and sticklebacks.

Though innovative and popular at the time, Huet's classification is far too simplistic to account for the diversity of river systems, both in their natural state and when substantially modified by humans. Many rivers, particularly those flowing over hard rock and steep topographies to the west of mainland Britain, run rapidly without any form of lowland zone. By contrast, many rivers and tributaries on lower and flatter land with slower flows, predominantly throughout the east of England, lack rapid reaches. Other smaller, well-vegetated channels, though lowland and meandering, may have insufficient volume to sustain species such as common bream. The differentiation of species suited to varied flow regimes is nevertheless valid, though river habitats are more generally now considered as a continuum: the faster and slower, and shallower and deeper reaches merging with each other, influenced by geology, topography, hydrology and often highly localised natural features as well as human modifications. This way of thinking about different flow- and sediment-related river characteristics also works for plants, water birds and other river life.

Perhaps as important as the changes in habitat along a river is the diversity of habitats within any particular reach. In their natural state, before pervasive human intervention, the European landscape was vastly more wooded before the mass felling of its once-extensive forests, attributed initially to our Iron Age forebears, then accelerated massively by Roman occupation clearing land for grazing of introduced sheep. Wildwoods covered hills, plains and valleys. Trees shaped the ever-changing courses of rivers, breaking them into braided networks as channels became choked by fallen wood, another formerly ubiquitous feature of riverscapes also significantly influenced by the activities of once-widespread beavers. Diverted water swept clear new channels, creating moist floodplains that may have been kept open by herds of grazing animals or else reverting back to climax wooded cover over longer timescales. This dynamic mosaic supported a diversity of wildlife, from the aurochs (progenitors of modern cattle) that grazed spaces scoured clear by spates and which became quickly vegetated by a profusion of grasses and low flowering plants through to forest-dwelling creatures of all types. Rivers were connected with extensive floodplains and wetlands. But 90% of Britain's wetland heritage has been lost since Roman times, and many floodplains too are converted for agriculture and urban and industrial development. Modern rivers are substantially depleted in habitat. This is particularly so for ecotonal habitats between the fully wet and the fully dry supporting the refuge, breeding and feeding needs of a diversity of fish and other groups of organisms.

Natural river features are further substantially modified across the British Isles by other human activities, including impoundment of flows by weirs and dams, channel straightening, bank reinforcement, as well as landscape drainage and flood embankments and other infrastructure that disconnects channels from floodplains. This complex juxtaposition of natural and human-created factors forms an often quite-small-scale mosaic of microhabitats. A common example of this is that brown trout and barbel may thrive along with minnows, dace, bullheads and chub in the faster, better-oxygenated waters over coarse gravel immediately below a weir on a lowland river whilst, perhaps as little as half a kilometre downstream, shoals of deeper-bodied fish such as common bream and roach inhabit the deeper, slower flows in water stilled over a silty bed above the next weir.

Then we have to consider the influence of marginal backwaters, vegetation in the channel and its edges, tributary streams and ditches and connected wetlands. This variety of habitat, where it has not been modified by urban, industrial and agricultural encroachment, can play host to a wider diversity of fishes including, for example, ten-spined sticklebacks and the spined loach that fare less well in open channels containing many competitors and potential predators.

## 6.3 STILL WATERS

The heterogeneity of rainfall, topography and geology across the British Isles also creates a wide diversity of standing waters. These range from large glacial lakes scoured out by the movement of glaciers during the last Ice Age through to lowland meres and small, weeded and sometimes temporary farm ponds.

Waters that dry out intermittently, whilst host to fish with resilient life stages (generally eggs) in many countries, tend to be fish-free in the British Isles. However, they may be important for plant, amphibian and invertebrate species adapted to the temporary presence of water but that are not resilient to the attentions of fishes. Insensitive 'rehabilitation' of temporary pools, including digging them out and stocking the now-permanent water with fish, can adversely affect their conservation value where it affects these other organisms. However, small, weeded ponds holding permanent water may be important habitat for species such as crucian carp and ten-spined stickleback that thrive in well-vegetated small water bodies often of low water quality, enabling them to evade larger, open water species with which they generally do not compete well. Silver bream, rudd, perch and roach can also thrive in small duck ponds, dew ponds, marshes and other enclosed water bodies; rudd and perch, in particular, often breed profusely, although frequently forming stunted populations where they outstrip food supplies.

**Temporary water bodies can be important for many adapted species, such as amphibians and various insects, the survival of which is substantially influenced by the absence of fish predation. (Image © Dr Mark Everard.)**

**A deep and cool upland lake, often providing habitat for glacial relic species.**

The large glacial lakes of northwest England, Wales, Scotland and Ireland provide deep, cool conditions suitable for scarcer fishes, including the 'glacial relic' species: Arctic charr and the whitefishes (vendace and European whitefish). These are fishes of deep, well-oxygenated, cool still waters and so are under considerable threat from nutrient enrichment and organic pollution in contemporary landscapes as these substances combine to depress the oxygen concentration in deeper and cooler lake layers. Introductions of fish such as ruffe and roach into these waters, in which they are not naturally found, are also threatening successful breeding of some whitefish and Arctic charr populations.

Particularly in England, many larger still water bodies are man-made rather than natural. These include, in particular, gravel and other quarrying pits and reservoirs. However, these can be productive fish habitats. Older gravel pits fed by groundwater mature over a period of decades into habitat-diverse waters that support mixed fish populations. Many reservoirs managed for both water storage and recreation, including angling, are stocked with brown trout and the non-native rainbow trout, recognising the importance of good water quality. They can also provide excellent habitat for large predatory fishes such as pike and zander, smaller predatory species such as perch and a host of other fish species that thrive in still water.

## 6.4 CANALS, DRAINS AND DITCHES

Canals and drains are a diverse set of waters, mixing features of rivers with those of still waters. One thing that all active canals share is simplified habitat, often with reinforced banks that are poor or lacking in marginal features. Actively navigated canals are subject to constant disturbance of sediment due to boat passage, clouding the water and uprooting most submerged vegetation.

**Gravel pits can mature into rich habitats for fish and other wildlife.**

Some canals are connected to rivers and have a good flow of water, with features and fishy inhabitants akin to the rivers with which they are linked. Others are slack, and many more abandoned, with features more in common with still waters. The canal network is populated by a wide range of fish species. The flowing waters of the Kennet and Avon Canal, well interconnected with the Kennet river in its eastern extent, has strongly flowing water in many stretches that support riverine species such as barbel and chub. Many other canals have predominantly still or sluggish flows, favoured by species such as common bream, silver bream, tench and common carp. Predatory fishes such as pike and perch thrive in canals and, where the water is turbid because of boat wash suspending sediment, zander can proliferate owing to their adaptations to hunting in low light. When canals become abandoned, barely or no longer used for navigation, they become progressively thickly vegetated and tend to infill with silt, but can provide rich habitats for water plants and birds, as well as fishes that favour still water and dense vegetation such as rudd, crucian carp, ten-spined stickleback and spined loach.

Large fenland drains and other drainage ditches are an enlarged version of this form of slow or still water habitat, some navigated but many not. Consequently, their fish fauna have many similarities with that of canals.

Ditches may often be overlooked. However, they are a unique still water habitat. Even ditches that dry in the summer may be important spawning habitats for pike and other vegetation-spawning species seeking secluded places to lay their eggs away from the expanse of the rivers and large still waters in which adult fish live out their lives. The profuse vegetation in these ditches can play host to a diversity of invertebrate and other small food items invaluable

Canals can be habitat-poor, though tend to diversify where rarely navigated or abandoned.

Fenland and other drains resemble canals in terms of the types and paucity of habitats they provide for fish and other wildlife. (Image © Dr Mark Everard.)

Ditch and backwater habitats can be important for smaller fishes but also for the refuge, nursery and feeding needs of many larger fish species. (Image © Dr Mark Everard.)

to the earliest life stages of fish. Ditches with permanent water can also be important for many species of invertebrates, water plants, amphibians and other organisms, including fish that thrive in still, well-vegetated conditions, including spined loach, three-spined sticklebacks and ten-spined sticklebacks.

Overzealous management of drains and ditches can be problematic for the wildlife they host. A routine management regime that leaves undisturbed reaches on an alternative management cycle can help preserve their ecological value whilst retaining their functional purposes.

## 6.5 TRANSITIONAL WATERS

As rivers meet the sea, they naturally broaden into wide estuaries, often with the formation of tidal mudflats and saltmarshes. Estuaries mark a transition from fully fresh to fully saline coastal water. For this reason, they are known as 'transitional waters' in European and some UK legislation. 'Transitional' is a useful term as the interface between fresh and salty is often hard to define, and highly variable with season and weather conditions. Saline water intrudes

Estuaries represent important 'transitional waters' between fresh and sea water, their naturally extensive and diverse habitats exploited by marine visitors as well as by migratory fishes adjusting osmoregulatory performance as they transition between saline and freshwater life stages. (Image © Dr Mark Everard.)

far up lower rivers when flows decline, whilst at other times strong flows may reduce the salinity of coastal waters for appreciable distances. Also, there is often a complex matrix of waters of different salinities within estuaries, not merely between different inlets, but with fresh water overlaying denser saline water where mixing is slow.

In the upper, fresher-water regions of this estuarine zone, fishes of the lower reaches of rivers that tolerate some salinity may be found. These include often dense shoals of common bream, roach and other deeper-bodied fishes, as well as predators such as pike. Further down into increasingly saline water, few fish species live solely in estuaries, though some such as the unrelated smelt and sand smelt and the flounder may commonly be found occupying estuaries. However, transitional waters are important habitats for fish species that move from their adult form in the sea to breed in fresh waters (so-called anadromous fishes, including species of salmon, trout, shad and lamprey) and those that live as adults in fresh waters but return to the sea to breed (catadromous fishes such as European eels). Then there are the opportunists that reap the rich feeding of estuaries and can penetrate a long way inland to feed generally during the summer, including the European seabass, flounders and three British species of mullet.

Some other species capable of survival in both fresh and saline waters (three-spined sticklebacks) pass through estuaries. In fact, along coastlines with steep and short rivers that spate and are inhospitable during the winter months, three-spined sticklebacks may be flushed out into coastal waters to overwinter; a familiar sight off the west coast of Scotland. Zander also are known to tolerate brackish water, using larger estuaries and low-salinity coastal waters to move between river systems.

Estuarine habitat is, however, under threat, predominantly through conversion for urban, port, marina, industrial, agricultural and recreational uses. Further pressures arise from factors such as pollution, water resources and invasive species. We will return to these issues in Chapter 11.

## 6.6 SUBTERRANEAN FISHES

Cave environments are poorly explored because of their inaccessibility, particularly those caves that are full of water. However, it is believed that there are some two hundred species of cave fish living in various parts of the world. Until 2015, no dedicated cave species were known in Europe, until the surprise discovery of a 'cave loach' by divers in the Danube-Aach system in southern Germany. This fish, as yet not formally classified by science but considered to be in the genus *Barbatula* (the same genus as the stone loach, *Barbatula barbatula*, but biologically distinct), is the most northern species of cave fish ever discovered. The loach is pale pink, scaleless with blood vessels showing through the skin when illuminated, has poor eyesight, and inhabits the large, pitch-black underground Danube-Aach cave system. It is thought that the species diverged from surface-dwelling fish as recently as 16,000–20,000 years ago, which is very recent in evolutionary terms.

There are, however, no known specialist cave species in British waters. Nevertheless, fish, amphibians and bats do occupy British cave habitats. Some of them do so accidentally, or else they use them for shelter, as there may not be enough food to sustain them. An exception here may be bullheads, which are notoriously loyal to their territories and are found in caves particularly in Wales, where they occur year-round. These bullheads do not seem to be affected by the dark, and their colour does not seem to change.

Blanched brown trout are also found in caves, though this is a temporary adaptation as their colour is rapidly restored when exposed to daylight. The colour of brown trout is highly adaptive to their surroundings, and those found in caves generally take on the pale colouration of surrounding rocks. Brown trout are quite capable of feeding in the dark, so this is not a problem

other than the general lack of food, suggesting that these active fish are more likely to be cave visitors than committed cave dwellers. Other freshwater fishes found occasionally in caves, more probably visiting than resident, include minnows and flounders, as well as European eels, Atlantic salmon and three-spined sticklebacks.

There are, however, many rivers running unsuspected beneath the feet of city dwellers. All settlements depended on local fresh water bodies for drinking, watering stock, cooking and washing and a host of uses including irrigation, defence, navigation and waste disposal. Yet, as cities sprawled, it was commonplace for them to pollute and develop over the very rivers that gave them life, depending instead on imports from ever more remote sources for daily needs. London, for example, rapidly converted its life-giving streams into informal sewers, paving them over as they morphed from sources of vitality into a filthy nuisance. In Mayfair, the Tyburn was culverted under mews whilst, in West Norwood, the Effra was entombed beneath new Victorian villas. The River Fleet, from which Fleet Street derives its name, was re-engineered as a Venetian-style canal by Christopher Wren after the Fire of London, yet quickly turned into a foul sewer receiving all manner of wastes from the burgeoning city, such that it was subsequently buried out of sight and mind. In the city of Bristol, the River Frome enters a culvert midway down what is now the M32 motorway to emerge into the floating harbour around 4 km downstream, with the bustling city built overhead. Many such 'lost' urban rivers remain memories in place names: in London, the Kye Burn once watering Kilburn, Bayard's Watering Hole now Bayswater, Peckham Rye meaning 'village by the River Peck' and so forth. Yet, with improved pollution control, many of these subterranean rivers can contain fish. When I was involved in major re-sewering and wastewater treatment projects in the northern English city of Bradford, locals told me that thriving brown trout populations inhabited the upper underground rivers above the city. Furthermore, there are many tales of resourceful Londoners fishing down drain gratings for eels thriving beneath the city street to supplement their food rations during the Second World War. The reality then may be that Britain's cities and their many 'lost' rivers may be a secret resource of freshwater fishes.

## 6.7 FISH MIGRATION

The life cycles of anadromous fishes – those living as adults at sea but returning to freshwater to breed (such as Atlantic salmon and shad species) – are well known. The life strategies of catadromous fishes – those living in fresh waters as adults but returning to sea to spawn (including the European eel) – are also commonly understood. However, virtually all of Britain's freshwater fishes, with the exception of some small species such as the spined loach and bullhead, exhibit some forms of migratory behaviour throughout their often complex life cycles and between seasons.

Even often-perceived sedentary species such as roach and dace can carry out prodigious upstream spawning migrations – distances of hundreds of miles have been recorded across their Eurasian ranges – in order to reach suitable spawning sites. Many other riverine fishes also tend to migrate upstream to spawn, in part to access suitable habitat (such as well-flushed gravels in the case of chub and barbel) that may differ from their ideal habitat for the remainder of the year, and perhaps also as an adaptation to counteract the tendency of river currents to displace them downstream, particularly during their juvenile stages when they are less able to resist strong flows. Weirs and other obstructions, both man-made and natural, that break down river connectivity can have significant effects on the fragmentation or even progressive loss of populations of the most vulnerable species.

However, migration laterally in freshwater ecosystems may also be important for fish behaviour and, particularly, for spawning success. Pike, for example, tend to access well-vegetated backwaters from rivers and lakes in order to spawn early in the year. Arctic charr also generally need to leave their deep-water lake environments to run tributary streams to spawn in early winter, whilst tench and other generally deeper-water species of still waters need to access warmer and well-vegetated margins to spawn successfully in early summer. Marginal habitat in both still and running waters is generally important for reproductive success, shallow and better-vegetated waters warming disproportionately in spring as well as providing refuge and feeding for juvenile stages of fish species that may subsequently develop to live out their adult lives in quite different habitats.

Areas of deep water, bays and inlets, such as tributary streams and ditches, and marginal stands of reeds and overhanging bankside vegetation may also provide crucial refuge during spate flows. Many fish species actively seek out these types of refuges at these times or may be washed into their slacker flows on a rising spate. Such deep-water and well-covered marginal habitat may also enable fish to evade their various bird, mammal and other fishy predators during daylight or in clear water. At many times of the year, it is also common for fishes retreating to the cover of deeper water by day to run up to more food-rich shallower waters at dusk and by night under the cover of darkness.

Adequate extent and diversity of habitat in brackish waters is also crucial for species migrating between saline and fresh water, providing for their diverse needs as their physiology adjusts as they move between differing salinity regimes.

The migratory needs of all freshwater fishes have to date very much been overlooked for many or most species. However, they are important considerations for locating fish throughout the year, for meeting their needs throughout their various life stages and between seasons, as also for the sustainable management of freshwater ecosystems.

## 6.8 SO HOW DID BRITAIN'S FRESHWATER FISHES GET THERE?

'Doggerland' may not be a land mass with which many people are familiar, but it has been hugely significant in the movement of species into what is now Great Britain (the largest of the British Isles, today politically comprising England, Scotland and Wales). Doggerland today lies beneath the southern part of the North Sea, its name reflected in the contemporary Dogger Bank after it was flooded by rising sea levels between 6,500 and 6,200 BCE. In its former state as terrestrial tundraland, Doggerland was home to prehistoric human settlements and connected Great Britain as a peninsula to continental Europe during and after the last glacial period until the middle Pleistocene. The Rhine, Scheldt and Thames river systems converged to form the Channel River, today the English Channel, carrying the combined river flow to the Atlantic. North of Doggerland, a massive ice lake was retained. As ice melted at the end of the Ice Age, sea levels rose and the land began to tilt as the massive weight of ice reduced. A catastrophic event resulted, in which the ice lake broke through as a megatsunami around 6,200 BCE, submerging Doggerland and cutting off what was formerly the British peninsula.

Aside from species with a marine life stage or tolerance of salinity, freshwater fishes cannot move between isolated catchments without assistance. Naturally, the spread of species between drainage basins would be limited by geographical barriers. Consequently, historic connections of British rivers to continental systems explain a great deal of the natural distribution of freshwater fish species across the British Isles. Barbel, for example, naturally occur across Europe northwards of the Pyrenees and in England from rivers draining into the Humber in the north down to the Thames in the south. Formerly absent from other regions, they are now widely introduced and naturalised in many English, Welsh and some Scottish rivers. *Giraldus Cambrensis* (or 'Gerald of Wales' c. 1146–c. 1223), a medieval clergyman and chronicler of his times, published an account of *The History and Topography of Ireland*, recording that '...*pike, perch, roach, gardon, gudgeon, minnow, loach, bullheads and verones...*' were absent from Ireland. Giraldus observed that all the Irish species of freshwater fish known to him could live in salt water, prominently brown trout, Atlantic salmon and Arctic charr amongst eleven migratory or brackish-tolerant species of freshwater fishes that include pollan, three-spined sticklebacks, European eels, smelt, shad, three species of lamprey and the increasingly rare common sturgeon. These fish were all able to colonise Ireland's freshwater systems without man's interference, unlike many other fish species well suited to Ireland's diverse fresh waters, such as bullheads, grayling and gudgeon. Following introduction, common bream, roach, dace and tench have spread and thrived in Irish waters.

In fact, freshwater fishes are probably the most heavily modified element of the British fauna, with many introductions for sport, food, ornament or by accident, such as 'stowaways' with other stocked species or in bilge water, or by invasion through human-made canals and water transfer schemes. The ramifications of some introductions are serious and are addressed elsewhere.

There is an urban myth that fish eggs get carried to new waters attached to vegetation adhering to the feet of ducks. Appealing though the idea may be, there is nevertheless no shred of scientific evidence that it actually occurs. Ducks in flight are streamlined beasts, structured not to drag loads of debris in their wake. Furthermore, the capacity of fish eggs to withstand such exposure has to be called into question. An additional popular myth is that fish can be dropped into new waters by careless predatory birds. Even were this likely, and fish (most likely injured) were able to withstand the impact of dropping from a great height into a suitable body of water, it would be a hugely inefficient strategy upon which to rely.

**Common Bream have thrived and become widespread since their introduction into the fresh waters of Ireland, from which they were formerly absent.**

Another more far-fetched sounding but possibly more likely explanation is 'rains of fish'. Many reported events of 'raining fish' around the world, from Egyptian times and from as far afield as Honduras and India, have been explained as underground river systems surcharging to the surface and stranding small fishes on land. However, well-documented reports of other events suggest that it can in fact rain fish. One such event occurred in Knighton in the Welsh county of Powys at 2:45 p.m. on Wednesday, 18 August 2004, a rain of fish falling with a thunderstorm and heavy shower a good 50 miles from the sea. Eyewitness reports describe the fish as being 'like minnows', some of which were still showing signs of life when they landed. Plenty of explanations had been advanced, such as reports of fish being observed being sucked up from the sea by tornados. Another 'flying fish' incident reported in London in 1984, in which the specimen was identified as a six-inch flounder, was attributed to it being sucked up from the neighbouring River Thames in a waterspout. A shower of two-inch (5 cm) fish was also reported on 6 August 2000 in Great Yarmouth in the English country of Norfolk. Are these reports of rains of fish credible in themselves and also a feasible means for movement of fishes? More plausible are extreme flooding events during which formerly isolated catchments can become temporarily connected, particularly across close headwaters or when large flushes of fresh water substantially dilute estuaries and coastal water. What is certain is that any natural process of migration of southerly species into regions of Britain formerly glaciated would be very slow and certainly massively slower than the rates at which humans are moving them around for better or for worse.

# SEX LIVES OF THE BRITISH FRESHWATER FISHES

## 7

One of the imperatives of evolutionary success is survival to pass on genes to the next generation. Britain's freshwater fishes exhibit a range of strategies evolved to feed, evade predation and disease and also to breed in the challenging environments they inhabit. In this chapter, we look at the reproductive strategies of these fishes.

Animals in general exhibit a spectrum of reproductive strategies. At one end of the scale are those that produce few fertilised eggs or young and frequently accord them great care in upbringing. At the other is the *high fecundity* (or, in other words, more prolific) reproductive strategy. As a generalisation, many of Britain's freshwater fishes tend towards the high-fecundity end of the scale, though some groups exhibit parental care.

To introduce the topic, the reproductive cycle of the roach – a high fecundity strategist – is explored in detail. This is not merely as I have a particular affection for this species. It also exemplifies some of the challenges common to reproduction in other species, many of which have quite different reproductive strategies.

## 7.1 REPRODUCTION IN THE ROACH

In ecological terms, roach are something of generalists. They have a wide diet, adapting to seasonal availability of food, and are also tolerant of a range of flow conditions and even of low salinity in upper estuaries. These attributes enable them to exploit a wide array to natural conditions in both flowing and standing fresh waters. Consistent with this generalist lifestyle, roach also have an extraordinarily high fecundity, offset by a correspondingly high mortality. To provide the young with a kick-start in life, however, spawning is timed to coincide with the spring explosion in food availability.

**The roach is a generalist species, the reproductive behaviour of which is used here as illustrative of general patterns in cyprinid reproduction.**

### 7.1.1 Preparation for spawning

So important is spawning that preparation for it, after roach mature at three to four years old, is almost a year-round occupation. The bulk of the summer season is spent feeding on the seasonal abundance of invertebrate and other food, recovering from the stressful spring spawning season. As early as July, when fully recovered, roach and many other cyprinid fishes begin to develop new sexual organs (more usually referred to as *gonads*), which start to increase significantly in weight throughout the late summer.

In mature female roach, a series of hormonal triggers (significantly including *gonadotrophin* released from the brain as well as *oestradiol*) stimulates the production of the egg yolk protein vitellogenin, which is transported by the blood stream for incorporation into the eggs. Yolk production and the storage of food reserves in the liver and muscles occurs throughout the winter, the eggs continuing to mature during the winter and increasing in rate of maturation as the spring approaches. Roach accordingly broaden in girth, particularly noticeable at the 'shoulder' (between head and back) and the depth of the belly. Female Roach become notably plumper, with fully mature spawn usually accounting for 15% of the body weight of a healthy female roach and accounting for around 19% of their annual conversion of energy into body tissue. As early as February or March, male roach develop hardened tubercles on the head and shoulders, shedding mucus such that they become rough to the touch. Female roach do not develop these tubercles. This roughness is thought to assist the males in adhering to the females they chase during the spawning act and may stimulate the release of eggs.

### 7.1.2 Spawning migration

Roach, like many fish species, tend to migrate upstream as spawning approaches. In rivers, this is common with the roach reaching streamy water, important for the subsequent

Tubercles on the head of a mature male common bream in readiness for spawning.

oxygenation of eggs but also perhaps compensating for the general downstream drift of juveniles with the current. In still waters, the fish may move into tributary streams, or else towards banks experiencing the most wind-generated turbulence. Roach do not spawn in estuaries, the spawning migration also serving to bring them upstream into fully fresh water to lay their eggs.

Scientific studies have found that around 90% of spawning roach home in on the exact same spawning sites used each year, discrete spawning groups segregating themselves from the large shoals they associated with for the rest of the year, potentially retaining genetically distinct strains of roach even within large and apparently mixed populations. In continental Europe, roach have been found to migrate for hundreds of kilometres up rivers each year to spawn.

### 7.1.3 Spawning

Roach spawn in late spring, anytime from April to the end of June in exceptional years and depending on latitude and temperature. The most significant environmental trigger to induce spawning in roach is day length (known as *photoperiod*), as it is this factor that stimulates production of the hormone gonadotrophin from the brain's light-sensitive pineal gland. Temperature is of secondary importance as an environmental trigger, fine-tuning the exact timing of spawning. Spawning is thus timed for when the water has warmed significantly, water plants are established and small food items are available for the emerging fry which require immediate and adequate nutrition to grow rapidly.

Roach are social spawners, spawning in large shoals in which male and female fish mix freely. Roach can spawn on a variety of different substrates ranging from submerged plants,

Roach spawning communally on a spawning board installed in the Hampshire Avon by the Avon Roach Project, as part of a long-term mission to enhance the river's formerly declining roach population. (Image © Avon Roach Project.)

the underwater stems of emergent plants or on woody or rocky substrates (so-called *plastic phytolithophils*). However, the favoured spawning substrate of roach is vegetation in shallow, well-oxygenated water, with moss-encrusted weir sluices or underwater tree roots favoured in rivers. In still waters or where this habitat is sparse or lacking, roach will readily spawn on the submerged shoots of reeds, underwater tree roots or overhanging terrestrial vegetation. In canals, roach often prefer to spawn on submerged vegetation such as Canadian pondweed or on filamentous algae, usually close to the water's surface.

In preparation for the spawning act, shoals tighten up near the chosen substrate and the male roach start to swim in irregular circles, females zig-zagging through the vegetation with one or two males adhering to their sides. The females shimmy slowly with rapid tail beats as sticky eggs are released, fertilised by attendant males that tremble as their milt is released to mix with the eggs as they float in the water. At maturity, a female roach usually produces 1,000 to 15,000 eggs, each about 1 mm in diameter, but a larger female of around 1 kg is capable of producing up to 100,000 eggs. Typically, 98%–99% of the sticky eggs are usually fertilised, most of them adhering to the chosen spawning material.

Most roach survive the spawning period, though the event is stressful and there tends to be significant mortality, particularly of old fish. A period of recuperation is essential, with many fish needing to recover from split fins, lesions or lost scales on the flanks, and to restore depleted bodily reserves. At this time, roach often gravitate to streamier sections of rivers or the most turbulent sections of lakes, the established wisdom being that they do so to 'clean up' (though this is not scientifically proven and it could be that they simply take time to disperse from their spawning sites).

### 7.1.4 Eggs, larvae, fry and onward growth

Roach exhibit no parental care of eggs, which are subject to massive mortality through predation by fish, amphibians and invertebrates; fungal attack; stranding as water levels drop; wash from boats; and many other causes. For surviving eggs, most of the volume of which comprises fat-rich yolk, the embryo develops rapidly. Some of this rich food remains attached to the fish as a yolk sac after hatching. Under typical British conditions, varying significantly with temperature, roach eggs hatch between four and ten days, though they can take as long as thirty days in the unusually low temperature of 7°C (45°F).

Roach hatchlings are very small, only some 4.5–6.5 mm (about a fifth of an inch) long, far shorter than a human eyelash and about the same width. The hatchlings are known as larvae, or 'external embryos', as they are far from completely developed and still attached to a residual yolk sac that they feed upon for the first few days. The larvae lack skin pigment, appearing as little more than slivers of glass to the naked eye. These early juveniles avoid light (negatively

Eggs of the dace, typical of those
of other cyprinid fishes, including
roach.

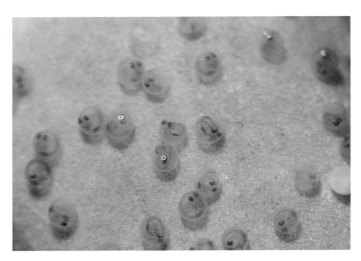

phototactic), helping them stay close to vegetation and other cover, avoiding at least some of
their predators. As they consume the remaining yolk, typically for some two to five days, the
larvae hang on to the spawning substrate using specialised adhesive glands on their heads.

Once the yolk is fully consumed, the juvenile roach detach from the sites to which they
had been resting, starting to swim and feed freely. The young at this stage, from the free-
swimming stage when the body straightens until the bones are fully hardened (ossified) and
the fins are fully differentiated, are known as *larvae*. However, the fins and gut are still only
poorly developed at this stage, providing limited ability to move and avoid their many predators.
Their small size at this stage, generally only around 7 mm long, restricts the diet substantially.

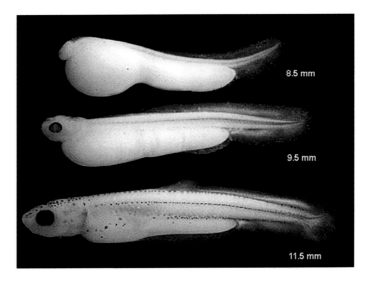

Larval development stages in
barbel with yolk sacs in different
stages of absorption, the lower
image with mouth and dorsal fin
starting to form, enabling the
juvenile fish to feed externally.
(Image © Dr A. C. Pinder.)

The first food items are usually individual algal cells, though near-microscopic invertebrates such as rotifers become increasingly important in the early diet. As noted previously, juvenile fish often have far bigger eyes than adult fish in proportion to their bodies, reflecting in part a greater reliance on locating small, mobile food by sight, with feeding virtually ceasing at night. Incomplete development of the digestive tracts of roach larvae creates a further limitation on effective feeding, the larvae needing to ingest approximately 50% of their body weight each day, growing rapidly throughout the first few weeks of free-swimming life. It is common to have 95% larval mortality.

The bones progressively ossify over the first few months, and the fins and gut become fully formed, after which the juvenile roach become known as fry. Rather than shying away from light, fry become *positively phototactic* (attracted to light), reflecting changes in the diet and life habit. Fry continue to feed predominantly by day, but also begin to detect and feed on small planktonic animals at night using their lateral line systems. As the fry increase in size, they begin to outgrow their plankton diet and switch rapidly to larger benthic fare, though large water fleas in the plankton remain an important component of the diet throughout the life of the roach. A peak growth rate of 30% of body weight per day occurs during their first summer, a yearling fish typically attaining a length of 5–7 cm (2–2½ inches). By the third year, faster-growing roach may attain a length of 10–15 cm (4–6 inches). After three or four years, roach mature sexually and join the spawning shoals.

Subsequent growth is linked to seasonal cycles of temperature and food availability. The metabolism and growth rate of a roach is highest in the summer, but growth diminishes

**Cyprinid fry, such as this juvenile bleak, start life around the size of a human eyelash.**

dramatically or ceases altogether during cool winter months both because of food availability and the diversion of resources into preparation for spawning. One consequence of this well-defined cycle is that the rate of growth of scales, which like other bony parts grow by accretion of new bone on the outer edge, gives rise to concentric growth rings that provide an indication of the age of a fish without need to kill and dissect it. Fish scales consequently are read very much like growth rings in a tree trunk, each growth ring (or *annulus*) signalling the end of a year's growth. When reading a scale, each annulus records a completed year of growth (though caution is needed as a period of adverse weather can also halt growth), so a fish that has completed two years of growth (i.e. two annuli are present on the scale) and is now growing in its third year is known as a 2+ fish.

### 7.1.5 Seasonal effects

The quantity of yolk in the eggs is crucial for giving the juvenile fish a rapid start in life, growing big and strong enough to evade predation and endure autumnal spates after their first summer's life. The amount of yolk laid down in eggs by female fish is highly dependent upon the conditions prevalent during the previous summer, autumn and particularly the winter. In good feeding years that are not too cold, adult fish recover quickly from their spawning exertions and have more resources available to channel into the production of the next year's eggs.

Climatic conditions in the juvenile's first summer are also vitally important. If cold or flood conditions prevail, low temperatures reduce metabolism and growth rate and food may also be limiting. These factors may result in substantial direct mortality. More crucially, that year's generation (or year class) of roach may be too small or otherwise ill-equipped to survive spate flows in autumn and other stresses throughout the coming winter.

Poor recruitment (addition of new fish to the population) can ripple through the age structure of roach populations over time. The number of fish in each year class of roach naturally diminishes gradually owing to mortality, resulting in healthy populations being pyramidal in structure – many more small fish and few large ones – and so divergence from this pattern suggests some problem with the environment.

### 7.2 REPRODUCTION IN THE OTHER CYPRINID FISHES

The life cycles of other cyprinid species is broadly similar to that observed in roach, though with some variation. All, with the exception of some invasive species that we will turn to later in this book, have a similar yearly cycle, though minnows and gudgeon may have extended breeding periods with multiple spawnings in a year. Most migrate to some degree, still-water species such as tench and common carp doing so into shallow, well-vegetated margins that warm quickly. All are communal spawners, though male common bream are observed to have some territorial behaviour around the best habitat within the bigger spawning shoal.

**Chub tend to spawn on well-flushed gravels in the late spring or early summer. (Image © Dr Mark Everard.)**

However, not all cyprinids spawn on vegetation. Barbel, chub and dace spawn on gravel in well-flushed, generally shallow reaches of rivers. Timing of spawning also varies significantly. Dace are one of the first species to spawn, sometimes as early as February or March. Dace growth also tends to be more continuous throughout the year in this cold-adapted species, making age interpretations through scale reading difficult. By contrast, fish species at the very north of their geographical range in the British Isles, particularly barbel, chub and common carp, may delay spawning until early summer when the water is warmest: June or even July are not unknown, and spawning may not even occur in cool years.

Barbel and chub are particularly susceptible to the seasonal effect of poor recruitment. Barbel often fail entirely to recruit in cooler summers. Chub populations show marked differences in year classes within populations, with 'good recruitment years' often occurring in drought or hot, dry summers with few fish recruited in intervening years.

One of the consequences of the group spawning behaviour of cyprinid fishes is that hybridisation between closely related species can occur. This is particularly the case where spawning habitat is sparse, such as in feature-poor reservoirs, resulting in competition amongst species for the same limited substrates. Hybrids between leuciscine species and some of the carps are particularly common, described already in considering those groups of fishes, but also addressed later in Chapter 11 as a conservation concern.

The marine and freshwater life cycle of the Atlantic salmon is complex. It is also well known, so this is where we will start in considering reproduction in the salmonid fishes as a group. However, rather less widely understood is the amazing flexibility that Atlantic salmon have evolved, apparently as 'insurance policies' for survival in adverse conditions. We visit these adaptations after considering the generalities of Atlantic salmon reproduction and before then turning to other salmonid fishes.

### 7.3.1 The basic life cycle of the Atlantic salmon

Almost everyone knows that Atlantic salmon run up rivers from the sea to spawn in their natal rivers, making prodigious leaps up any weirs and waterfalls that may lie in their way. This is the stuff of drama and inspiration. It is what is known as an anadromous life cycle, in which adult fish inhabit the sea but return to rivers to breed. In this section, we will look in a little more detail at the wider context of this part of the Atlantic salmon's life cycle. Later on, we will look at how salmon not infrequently break the 'rules' of this story as an adaptation for survival.

**Atlantic salmon migrating upriver famously leap over obstacles.**

Atlantic salmon eggs, deposited into depressions known as 'redds' in the winter in well-flushed open gravels in the cool and clean upper reaches of rivers, develop slowly to hatch during early spring. For the early days, these hatchlings are known as alevins, still attached to their yolk sac on which they feed for a number of weeks whilst remaining protected in the redd. After the yolk is consumed, the young fish become more mobile, emerging (or 'swimming up') from the gravel around April or May as fry. The fry are small, typically around 2.5 cm (one inch) long, and start immediately to feed on small river invertebrates. Atlantic salmon fry grow quickly, developing into well-pigmented parr characterised by between eight and twelve dark blue-violet 'dirty thumbprint' markings, known as parr marks, along their flanks. The parr are voracious feeders. Although they tend to be gregarious, and can be found at considerable densities on suitable gravel riffles and runs where river flows are strong, they are also territorial to maximise their individual opportunities to feed. Consequently, populations of salmon parr in rivers are 'density-dependent', or, in other words, limited by both the availability of suitable riffle habitat and the territorial space taken by each parr. The availability of suitable habitat is therefore an important limiting factor to salmon populations in this early life phase.

The parr life stage is of highly variable duration across the natural range of Atlantic salmon, depending on numerous environmental variables including water temperature, the suitability of habitat and food availability. During this time, the young fish are vulnerable to a wide range of predators, including birds such as grey herons (*Ardea cinerea*), fish such as pike and mammals including Eurasian otter (*Lutra lutra*) and mink (*Neovison vison*). However, typically after one to three years but as much as six in some cold, upland streams, the parr begin the process of smoltification, or metamorphosis into smolts. Smoltification entails the fish increasing their hypo-osmoregulatory performance (capacity to pump salts across the gill interface to excrete

**Alevins of Atlantic salmon lie in their spawning redds as the absorb their attached yolk sack.**

The parr stage of the Atlantic salmon is territorial, using gravel riffles as habitat from which to intercept passing small animal food items.

excessive concentrations from the blood into surrounding sea water) and developing a flush of silver scales. Once fully metamorphosed, the smolts are then able to inhabit saline waters, and they move downstream to enter estuaries. Generally, smolts remain in or around estuaries for some time, adjusting to higher salinity water. At this stage, they are extremely vulnerable to pollution or poor water quality, excessive predation or commercial fishery activities and other stresses such as diseases and parasites arising from fish farms. When suitably adjusted, the smolts head out to sea to begin their adult marine life phase.

As the smolts migrate to and occupy open-sea feeding areas, they demonstrate schooling behaviour. There is some evidence that recently spawned adults (kelts) returning to the sea may act as guides to these juveniles. The young salmon continue to grow and mature into adult salmon, feeding voraciously on a variety of crustaceans such as krill and shrimps, squid and other molluscs, other invertebrate prey and progressively larger fishes such as herring, alewives, smelts, capelin, small mackerel, sand launce (sand eels), blue whiting and small cod. Growth is rapid, the marine diet considerably richer than that of their natal rivers. Atlantic salmon breeding in British waters are found in a range of marine areas, the best known of which are the Norwegian Sea and the fertile waters of the continental plate off southwest Greenland.

Atlantic salmon typically spend between one and four years at sea before beginning their spawning migration back into fresh waters. A combination of behavioural and genetic factors trigger this urge to migrate, though the scent of natal rivers is a powerful signal. Single sea winter fish, known as grilse, tend to travel the shortest distances, not venturing beyond the Faroe Islands and the southern Norwegian Sea. However, those Atlantic salmon remaining at sea for more than one winter undertake longer migrations. This variable duration that individual

Atlantic salmon parr metamorphose into smolts, taking on a silvery sheen and becoming able to tolerate the transition from fresh to saline water.

salmon from the same year class spend at sea and their occupation of different areas of the sea appear to be adaptations to improve the security of the species should catastrophic events affect the marine or freshwater phases.

On entering river systems, Atlantic salmon cease to feed. From this point, they mobilise reserves of fat stored in their body tissues and can subsist on them for a year and sometimes more before eventually spawning in the gravel-bedded headwaters in which they hatched. For much of their time in fresh waters, mature Atlantic salmon tend to remain relatively dormant. Larger 'springer' fish have typically spent more than one winter at sea, entering rivers earlier in the year. The smaller grilse, staying at sea for a single sea winter, are typically late-running. In northern Russia, some Atlantic salmon are known to enter rivers on the Kola Peninsula in September of the year preceding that in which they will eventually breed. A small proportion of adult Atlantic salmon tend to stray to other rivers and breed within different populations, maintaining genetic diversity within populations.

During their relatively quiescent time in fresh water, be that in a river, loch or lough, the appearance of Atlantic salmon changes significantly. Male fish develop a powerful kype, or hooked lower jaw, which is used as a weapon to defend spawning sites later in the year. Atlantic salmon also grow progressively thinner as they mobilise their internal fat reserves, also losing their silvery guanine body colour and becoming greenish or reddish brown, mottled with red or orange. This colour change results from a build-up of carotenoid pigments derived from the invertebrates upon which they fed at sea and that are subsequently laid down in body fat. The pigments accumulate as a waste product when fat reserves are remobilised. The change in colour is particularly marked in cock (male) fish, the pigments within hen (female)

Cock Atlantic salmon develop a kype, a hooked lower jaw, used as a weapon to defend spawning territories against rivals.

fish being remobilised with fat reserves into the production of yolk in their maturing eggs. Both the appearance and the taste of the flesh changes with length of time spent in fresh water. Because of this factor, most salmon caught commercially for human food are intercepted at sea, in estuaries or at the bottom of river systems, where they are in prime condition.

The dramatic and often prodigious leaps made by Atlantic salmon as they run rivers have already been alluded to, adding to the iconic status of these 'kings of the fishes'. A peerless description appears in Izaak Walton's classic 1653 book *The Compleat Angler*:

> *…they will force themselves through floodgates, or over weirs, or hedges, or stops in the water, even to a height beyond common belief. Gesner speaks of such places as are known to be above eight feet high above water. And our Camden mentions, in his Britannia, the like wonder to be in Pembrokeshire, where the river Tivy falls into the sea; and that the fall is so downright, and so high, that the people stand and wonder at the strength and sleight by which they see the Salmon use to get out of the sea into the said river; and the manner and height of the place is so notable, that it is known, far, by the name of the Salmon-leap.*

Timing of spawning varies significantly amongst river systems, influenced by a range of factors including water temperature, day length (influencing hormonal triggers) and pulses of water down the river. Large spates can be important migratory triggers, encouraging and enabling migratory runs through or over significant obstacles. Generally, Atlantic salmon spawn between November and December. At this time, male salmon have taken up position in suitable well-flushed coarse spawning gravels, vying with and driving away competing cock fish as well as other species for the best territories. With territories established, the cock fish then entices ripe hen salmon. When the females are ready to release their eggs, they excavate 'redds' in the gravel bed using powerful shimmying body movements into which eggs are deposited, cock fish shedding their milt to fertilise them. When spawning has been completed, the hen fish covers the redd using her body to mobilise gravels from upstream. Often, multiple releases of eggs creating new redds can result in mobilised gravel covering previous redds already containing eggs. Spawning continues until the female is spent. In general, Atlantic salmon produce about 1,750 eggs for each kilogramme of body weight; a large female salmon of 10 km (22 lb) may therefore lay up to 17,500 eggs in one or more redds. As salmon eggs are heavier than water, they drop into cavities in the coarse gravel of the redd, though many eggs are lost at this time to predators including invertebrates and smaller fishes including salmon parr.

After spawning, the spent fish are known as kelts. Kelts can usually be distinguished by their thin shape, distension of the vent from which eggs and milt were released and parasitic flukes (often called 'gill maggots') on the gill filaments. Kelts also tend to be very weak with injuries from territorial fights, the attentions of would-be predators or sustained when passing obstructions on their travels upriver. Kelts are therefore extremely susceptible to secondary infections, particularly from *Saprolegnia* fungus, a high proportion of British Atlantic salmon succumbing after their first spawning. The change in body shape particularly with growth of the kype and exertions of fighting competing males for the best habitat for redd-digging, added to the direct rigours of spawning, are cumulatively often fatal for cock Atlantic salmon. However, the predominantly female survivors regain their vitality, recovering their silvery appearance and returning to sea to feed and rebuild condition before the next spawning run.

Eggs lying in the redd, flushed by cool and oxygen-rich water, remain dormant in the gravel throughout the colder months before hatching in the spring. The duration between egg fertilisation and hatching is strongly dependent on temperature. Egg and fry development and growth is directly related to 'degree-days' (mean temperature in °C multiplied by number of days). Eggs of the Atlantic salmon take approximately 220–250 degree-days to 'eye-up' (the stage at which a distinct eye is visible through the egg shell) and a further 220–250 degree-days to hatch.

### 7.3.2 Further remarkable reproductive adaptations in the Atlantic salmon

The initial overview of the Atlantic salmon life cycle illustrates what is typically and generally known about the species and also principles common to many other salmonids. However, there are some further remarkable adaptations in this life cycle that provide Atlantic salmon with flexibility not only to survive but to prosper through adversities and even over geological timescales.

Firstly, as already seen, adult Atlantic salmon can spend anything from one to four winters at sea, occupying different marine regions during this time. This means that, if adverse river or sea conditions result in substantial mortality, differences in duration at sea enable individual fish from the same year class to return to contribute to future generations.

Secondly, there is the timing of arrival and variable duration that Atlantic Salmon spend in fresh waters. Big 'springer' fish tend to be those that have spent multiple winters at sea, often requiring their extra strength to force their way upstream in more tempestuous flows and over larger barriers before river levels decline as the summer progresses. Later in the year, the grilse run is dominated by single sea winter fish. As an extreme example, a regional adaptation observed in the Russian stock of Atlantic salmon in the Kola Peninsula is that fish may enter rivers over a year before spawning, so that they are ready to run the more challenging reaches of river overcoming obstacles in high winter flows to reach spawning headwaters. This variability in timing and duration of entry into fresh waters is another 'insurance policy', ensuring that spawning fish reach the headwaters.

Thirdly, we have the phenomenon of 'straying'. Though, famously, Atlantic salmon home in using the scent of their natal headwaters imprinted upon them during the parr stage, a significant but small proportion of adult salmon enter other river systems. This not only ensures a degree of genetic cross-fertilisation but also enables salmon to colonise waters in which they are absent either naturally or because of catastrophe. This is illustrated by the remarkable pervasion of Atlantic salmon southwards to a latitude of about 40°N at both the eastern and western seaboards of the Atlantic Ocean. Interesting questions arises as to where these salmon got to throughout the Ice Ages and how they appeared again in regions previously covered in glacial ice. The southern European extent of Atlantic salmon is in the Gave d'Oloron draining from the Pyrenees and considered France's finest salmon river, twenty-six rivers

**'Springer' Atlantic salmon that have been at sea for multiple sea winters tend to run rivers earlier in the year and to be far larger than grilse, or single sea winter returning fish.**

draining the Cantabrian range and the Galician Coast on the Atlantic seaboard of northern Spain, and the Rio Minho and the Rio Lima rivers in northern Portugal which both rise in higher ground in northern Spain. These Iberian and southern French populations may have been of great significance for the survival of salmon during the last Ice Age ending some 11,700 years ago. Genetic studies suggest that current northern European salmon populations derive from just two distinct stocks, one believed to have survived extensive European ice cover in a southeastern ice lake refuge and the other occupying southern rivers remaining free from the advancing ice sheet. Similar refuge populations seem to have occurred in North America. The straying behaviour of a proportion of mature Atlantic salmon is the principal mechanism that has enabled the species to colonise new waters beyond their native streams, accounting for their current wide distribution.

Fourthly, there is the phenomenon of 'precocious parr'. Precocious parr are those that mature sexually though have not yet completed their transformation into sea-going adult fish. These small fish, perhaps less conspicuous as competitors to adult territorial cock fish, join the adults in their breeding frenzy. The eggs from a female fish can therefore be fertilised by an adult cock fish and several parr from a completely different year class, thus providing the maximum possible opportunity for successful spawning and the maintenance of genetic diversity.

A further strategy is seen in the case of landlocked populations of the Sebago Salmon (*Salmo salar sebago*), a sub-species of the Atlantic salmon found in four of the US Maine lake systems: Sebago, Green, Sebec and Grand. Native landlocked populations are also said to exist in eastern Canada, Scandinavia and eastern Russia. Though no such landlocked salmon population is known in Britain, this type of life strategy adds further flexibility to the species, appearing similar to that of the European southeastern ice lake refuge population that contributed substantially to seeing the Atlantic salmon through the last Ice Age.

The life cycle of the Atlantic salmon, though often told in simple terms, is in reality far more flexible, adapting to local conditions with many variants and providing a high degree of resilience in the face of changing and uncertain conditions.

### 7.3.3 Life cycles of other British salmonids
The sea trout, the sea-going form of the brown trout, has a similar sea-going life cycle to the Atlantic salmon, running rivers to breed in cool, well-oxygenated headwaters. Most British sea trout populations comprise fish that have spent only one winter at sea. However, some rivers, particularly in southern England, are known for their run of large sea trout.

The life cycle of the brown trout is in other respects similar to that of the Atlantic salmon, albeit without a marine adult phase, mature breeding trout running up rivers to spawn in cool,

**The sea trout is the sea-going form of the brown trout.**

well-flushed headwaters. Lake-dwelling trout run tributary rivers or, exceptionally, breed in gravel margins well flushed by wind-driven turbulence. Male trout guard territories over suitable gravels in the autumn or early winter, gravid female fish cutting one or more redds and depositing their non-sticky eggs. The eggs fall into the gravel's pores, to hatch in late winter or early spring.

Alevins emerge from the eggs in the early spring with yolk sacs still attached, swimming up as fry when the yolk is consumed. Fry and parr feed on aquatic and terrestrial invertebrates and small fishes. The general 'rule of thumb' of 1,750 eggs per kilogramme of body weight observed for female Atlantic salmon seems to apply for all British salmonid species.

Quite what differentiates brown trout from sea trout remains a mystery, though the triggers appear to be a mix of genetic and environmental. Populations of trout in steep, nutrient-poor catchments such as many smaller river systems on the west coast of Scotland, where food is sparse and spate conditions violent, may cause the overwhelming majority to take to sea. Conversely, brown trout populations in placid lowland rivers with better-buffered flows and richer feeding tend to produce lower proportions of migratory sea trout.

The life cycles of rainbow trout and Arctic charr broadly mirror those of the brown trout, including their sea-going strains, though none of these occur naturally in Britain. European whitefish and vendace too share a similar life history, emerging from deep, cool strata of glacial lakes to breed during the winter on marginal gravels around lake shorelines when the water is about 6°C.

Of the British salmonids, grayling differ by spawning in rivers in the spring, more or less around the same time between March and early May as the cyprinid species. It is for this reason

The grayling is a salmonid fish but spawns in the spring, so it is classified as a 'coarse' fish.

that they are classified as 'coarse fish' for angling purposes, despite their salmonid heritage. As spawning time approaches, grayling assemble communally on suitable, better-flushed gravels. Male grayling at this time darken and intensify in colour and become very aggressive towards other males in defence of discrete territories within the communal shoal. They hold their spectacular, iridescent blue-red dorsal and pelvic fins erect to display to females and to entice them to dig redds and lay their eggs. Male fish also use their sail-like dorsal fins to hold a willing female close by, curling it towards her prior to egg laying. Spawning behaviour may sometimes be diurnal, male grayling leaving the spawning gravels at night but returning in the morning to reclaim a territory. Most spawning occurs as water temperatures peak in the early afternoon. Grayling redds dug by the hen fish in well-oxygenated river gravels tend to be shallow. Typically, eggs deposited in the redds hatch after three or four weeks, the alevins remaining in the redd until the yolk sac is consumed, then developing into fry that 'swim up' to live thereafter as free-swimming fish. The grayling is a fast-growing but short-lived species, maturing after three or four years.

## 7.4 PATERNAL BROOD CARE BY BRITISH FRESHWATER FISHES

Although most species of British freshwater fishes, certainly most of the cyprinid fishes (with exceptions noted in Section 11.3.2) and all of the salmonid fishes, exhibit no parental care after spawning, this is not true for all fish species. Famously, 'male pregnancy' is widely known amongst seahorses and pipefishes (family *Syngnathidae*), the female seahorse inserting her ovipositor into a brood pouch on the front of the male's body where the eggs are brooded for two to four weeks. However, a number of British freshwater species also exhibit forms of 'new man' paternal brood care behaviour.

Male sticklebacks, including both the three-spined stickleback and the ten-spined stickleback in British fresh waters, also brood their eggs and juveniles. In the spring and summer spawning season, from March through to August, male sticklebacks build hollow nests from algae and other vegetation, into which they attract a number of females to lay their eggs. The male stickleback fertilise the eggs and drive off the females, then caring for the eggs and hatching youngsters. The male three-spined stickleback at this time transforms from its predominantly drab and silvery body colour, taking on a deep green or blue hue with brilliant red on the underside of the head and belly, the eye becoming electric blue. Male ten-spined sticklebacks tend to retain their drab colours, as do the females of both species. Stickleback nests are rounded, often spherical constructions that the male fish painstakingly builds in shallow water using filamentous algae, larger water plants, twigs and detritus stuck together using a secretion from the kidney rich in the glue protein spigin. Finished stickleback nests have a hollow passage through the middle, into which female fish are coaxed by the male's characteristic zig-zag dance, inviting her to inspect, enter and lay her eggs, which the male then fertilises before driving her off. Each female deposits anything from one hundred to four hundred eggs, each 1.5–1.9 mm in diameter. A successful male might entice several female fish to lay their eggs in his nest, which may therefore contain anywhere up to one thousand eggs. The male stickleback then guards the territory and nest, often driving off other, often far larger species of fish, all the time watching out for yet more females to spawn with him. The male continues

**Male three-spined stickleback in bright spawning colours.**

**Male three-spined stickleback building a nest.**

**Male three-spined stickleback releasing glue secreted from the kidney.**

to fan oxygenated water over the eggs with his pectoral fins, removing dead or infertile eggs. Once the eggs hatch, and the fry become free-swimming, the male stickleback then herds and guards them for the first few days after they emerge. The fry subsequently disperse into cover to start their independent lives, feeding on small animals.

The bullhead is another common British river species in which male fish take charge of parental duties. Bullheads are notoriously territorial species, living largely solitary existences in faster streams and well-flushed lake margins, not infrequently living in the 'cave' under an individual stone or large piece of woody debris throughout their whole adult lives. Bullheads spawn in Britain typically between March and May, male fish displaying to generally smaller females living in adjacent territories. If the male bullhead succeeds in luring the female into his lair, he then coaxes her to deposit a clump of around one hundred sticky eggs, typically 2–2.5 mm in diameter and amber in colour, generally on the roof of his territorial cave. After the eggs are laid and fertilised, the male bullhead drives off the female and guards them until they hatch after three or four weeks and for a further ten or twelve days thereafter as the larvae consume remaining yolk. Bullhead fry then become free-living, immediately dispersing to evade the potential predatory attentions of their fathers, taking shelter under stones and feeding mainly on small invertebrates.

Other large, introduced species exhibiting male brood care include the zander and the wels catfish. Zander spawn in late spring or early summer when water temperatures exceed 12°C, laying sticky eggs on hard substrates that may include rocks, plant stems or underwater tree roots. There are also videos on the internet of zander using discarded car tyres as spawning substrates. Male fish clean the spawning substrate and guard the eggs and newly hatched fry, after which the fry are abandoned to feed on zooplankton and other invertebrates, progressing to a mainly piscivorous diet as they grow. Wels catfish breed in early summer in the vegetated margins of lakes and large rivers, laying eggs on mounds of leaf litter. These nests are guarded by the male catfish until the juveniles, which are superficially similar to tadpoles, emerge to grow rapidly on a variety of invertebrates progressing to a wholly carnivorous diet.

The small, introduced and invasive topmouth gudgeon and sunbleak are a further two species in which male fish demonstrate a degree of brood care. However, these fishes will be discussed subsequently in Chapter 11 when considering their generally adverse impacts on the conservation of British freshwater fishes.

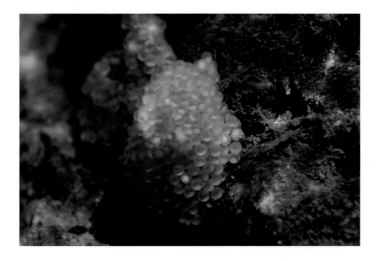

**Male bullheads entice females into their caves to lay eggs, generally on the roof, before fertilising them, driving the female fish off and brooding the clutch.**

## 7.5 THE LAMPREYS: LOVED TO DEATH

Reproduction in the lampreys is a fatal affair. After they have spent several years as ammocoete larvae in river sediments, metamorphosing then into their adult forms, river lamprey and sea lamprey migrate to take up a parasitic estuarine or coastal sea existence before returning to fresh waters to breed. The smaller brook lamprey has no such extended adult life stage, metamorphosing into an adult form that does not feed but exists purely to breed.

All three British lamprey species return to, or remain in, rivers to breed. Adaptations of lampreys to spawning reflect their early stage of evolutionary development. Lampreys spawn in the cool and clear headwater streams or cleaner gravels in main rivers, to which they migrate

**Brook lampreys often spawn communally shortly after metamorphosing, excavating depressions in river beds.**

**Sea lampreys collaborate in nest building before laying eggs and then dying.**

and where they build nests, or redds. Male and female fish collaborate in the building of these spawning pits, removing small rocks individually with their mouths and fanning smaller particles with their tails. Once the redd is ready, the adult fish deposit their sperm and eggs simultaneously with their bodies intertwined. Shortly after the eggs are laid and fertilised, the adults die. Ammocoetes, or larval lampreys, emerge from the eggs after one to two weeks, burrowing into the silt or mud of river margins where they feed by filtering microorganisms, algae and detritus from the water.

## 7.6 THE BITTERLING: PULLING A MUSSEL

Although the bitterling is an introduced species in British waters, it is a fish with one of the most fascinating breeding behaviours. Bitterling are small fishes favouring densely weeded regions of small lakes, ponds and slow-flowing streams with sandy or muddy bottoms, habitats also favoured by freshwater mussels. However, the relationship between bitterling and mussels is far deeper than just their similar habitat preferences. Male bitterling develop a brilliant spawning livery by April or May, the flanks becoming strikingly iridescent and the dorsal and anal fins taking on a bright red hue, with a triangular area each side of the snout bearing dense white tubercles. Female bitterling grow a fleshy ovipositor (egg tube) extending about 6 cm long from the genital opening.

Successful spawning depends entirely on several species of freshwater mussels. These mussels lie half-buried in the sediment, drawing water into their mantle cavities through a siphon tube that protrudes from the gaping halves of their shells and filtering plankton from the passing water. Male bitterling select a suitable mussel, which they then guard against other male fish. Once an obliging female bitterling has been attracted, she inserts her extended

Bitterling deposit their eggs within the mantle cavity of a freshwater mussel, the female bitterlings ovipositor tube accessing the mantle cavity through the mussel's syphon.

ovipositor into the mussel's feeding tube, sticking one or two eggs onto the gills in the space between the mussel's shells. The male then sheds his milt near the mussel, fertilising the eggs as the mussel draws water in over its gills. The same pair of fish may repeat the process, though male bitterling also tend to select other females to lay eggs in their territorial mussels. Female bitterling can deposit from forty to one hundred eggs, each approximately 3 mm in diameter. The eggs remain protected from predators within the mussel's mantle cavity until they hatch between two and three weeks later. Larvae initially remain within the protective mantle of the mussel whilst they absorb their yolk sac, before then metamorphosing into fry that leave the sanctuary of the host mussel. A further benefit arises from the mussel's response to declining water levels, these molluscs drawing themselves through the sediment into deeper water using their muscular 'foot'.

Mussels appear completely unharmed by this process. However, there is a symbiotic relationship as mussels release 'glochidia' larvae into the water, which latch onto the gills of fish to grow until they are ready to fall off and start independent lives on the bed of the pool or river. The tiny glochidia are not only protected within the fish, but benefit from dispersal. Returning to the bitterling, there is some *quid pro quo* in their spawning arrangements. Bitterling frequently act as hosts to the mussel's parasitic glochidia larval stage, which attaches to the gills.

Atlantic salmon and brown trout too have intimately evolved interdependencies with mussels, but in this instance, reproductive success of the mussels depends on the fish without reciprocal benefit. One of the most complex and dramatic such interrelationships is with the pearl mussel, a severely endangered mollusc found living in the bed of just a few exceptionally clean English, Welsh, Scottish and Irish rivers. Pearl mussels filter-feed from river water. Between June and July, male pearl mussels discharge their sperm into river water, which is inhaled by female mussels through their filter-feeding activities. Inside the mantle cavity of the female mussel, the sperm then fertilise eggs in special sacs adjacent to the gills. From these fertilised eggs, glochidia larvae are released into the water column in great densities, a single female pearl mussel releasing as many as four million larvae. The next stage of the life cycle depends upon the glochidia finding an Atlantic salmon parr or brown trout host. Once located, or more likely breathed in by chance, the hinged body of the glochidia larva snaps shut on the gills of the fish where it forms a cyst and continues to grow and metamorphose. Pearl Mussel larvae remain attached to the host fish until the following spring or early summer. At this point, they drop off as seed mussels, requiring clean, fine gravel in which they burrow to live out their long adult lives.

## 7.7 DRIFTING WITH THE CURRENT

Two species of freshwater fishes found in Britain leave their eggs to the mercy of currents, albeit both unsuccessfully in British waters.

The burbot, as we have seen, became extinct at some time in the 1950s or 1960s in British waters, despite lingering but unfailingly unsubstantiated rumours to the contrary. Reflecting their marine ancestry and close relationships with cod-like fishes, burbot spawn in the winter, often under ice. Mature burbot migrate into the shallow margins of large lakes and rivers, spawning communally at night on sand or gravel beds. The eggs are tiny, one to three million of them released by each female. Each egg contains an oil globule enabling it to float off into the water column. There is no brood care. Under this high-fecundity strategy, many of these eggs die. However, the few that survive hatch to produce minute burbot fry. The fry feed on small planktonic animals, progressing to larger invertebrate prey as they grow. Both temperature and flow regimes of the large, eastern-flowing English rivers from which burbot were formerly found, and their modifications under growing human development pressures, are likely factors in the demise of the British population.

Grass carp also release pelagic (open water) eggs that move freely with currents. These are naturally fish of large, turbid Asian rivers and associated floodplain lakes. Although the adult fish have a wide range of temperature tolerance, conditions for successful spawning include a relatively high temperature of 20°C–30°C, preferentially in shallow water. The eggs then drift in the current of large rivers, detained in linked pools and backwaters, with reports of juvenile grass carp travelling as far as 1,000 km from their original spawning grounds in their native region and some of the many countries in which the species has become naturalised after introduction. The fry too require a higher sustained temperature to that generally found in Britain. These factors combine to explain why grass carp fail to spawn in British waters.

## 7.8 THE MYSTERIOUS CASE OF THE EUROPEAN EEL

Many of us will have learned about the life cycle of the European eel when at school, or subsequently. It is quite a tale. This fish is catadromous, a mirror image of the anadromous fishes that spawn in freshwaters and live their adult lives at sea. European eels, or at least some of them, live out their adult lives in fresh water, returning to sea to breed.

I recall books when I was a boy with images of eels spawning in seaweed mats floating in the Sargasso Sea, near the Caribbean, after which larvae known as *leptocephali* spend some years drifting back to European shores on Gulf Stream currents. The juvenile eels at this point metamorphose into a small, transparent form known as 'glass eels', entering lower rivers. These tiny eels then work their way upstream and, as they do so, darken in body colour into 'elvers' that are basically tiny eels. The eels then move upriver, finding their way into pools, canals and all manner of other freshwater environments, where they remain potentially for ten or more years. At some point, a trigger stimulates a change in their form. As feeding adults, they are often referred to as 'yellow eels', characterised by small eyes, a slightly yellow patina over their drab, grey bodies and a large mouth for their predatory or omnivorous diet. The

spawning trigger sees them metamorphose into 'silver eels', far less slimy, with bigger eyes, a smaller mouth and an overall silvery tinge to the body. Triggered often by a new moon during an autumn spate, silver eels then migrate downstream en masse, some individuals known to move over land in wet grass to vacate enclosed pools and access rivers. As the Reverend William Houghton puts it in his 1879 book *British Fresh-Water Fishes*, '*Eels have the power of living a long time, out of water if the air is humid*'. Then, they move downstream to re-enter the sea, and it is generally assumed that they return to the Sargasso Sea to breed.

Well, they are the essentials of common knowledge, and some of it at least is true! However, as with the overview of the Atlantic salmon story above, this simple telling omits a lot of detail in a complicated tale and also makes several sweeping assumptions about elements for which there is simply no evidence or, even in some cases, where there is evidence to the contrary. But the life cycle of the eel is such a fascinating topic, as are various interesting tales surrounding it, that we will save the wider telling for chapter 10, 'The curious and curiouser world of the European eel'.

## 7.9 SEX LIVES OF THE REMAINING BRITISH FRESHWATER FISH SPECIES

The spawning habitat of pike has strong similarities with that of many vegetation-spawning cyprinid fishes, with two major distinctions. Firstly, large female pike tend to move individually into vegetated backwaters or shallows and are generally accompanied by several smaller male Pike. It is not uncommon for a female Pike to make a meal of some of the much smaller suitors that assemble around her vying to fertilise her eggs. Secondly, pike spawn very early in the year, from late February to May depending on conditions, such that their fry, which are also notably cannibalistic, are ready to feast on the fry of other and their own species spawning

**Larger female pike enter vegetated backwaters early in the year to spawn, accompanied by a number of smaller attendant male fish.**

later in the year. Sticky pike eggs adhere to vegetation, fertilised by more than one small male, receiving no subsequent parental care.

Species such as perch, ruffe and zander have a generally similar life history to that typical of the cyprinids, spawning in spring but preferring to do so on hard substrates such as submerged rocks and stones, woody debris and tree boughs or roots dipping or submerged in the water. Perch and ruffe eggs tend to be laid communally in short, sticky strings, the eggs and fry receiving no parental care. The juveniles of both species feed predominantly on small invertebrates, progressing to a diet of larger invertebrates, amphibians and fish as they grow. Zander share these general traits but, as noted previously, male fish exhibit a degree of guardianship over the eggs.

Stone loach and spined loach too follow this general pattern of spring spawning, both species shedding clusters of sticky eggs. Stone loach do so amongst gravel, submerged stones and plants, with some reports of female loach sometimes guarding the eggs. Spined loach deposit unguarded eggs on submerged plants, roots or stones. The fry of both species are very small, requiring dense cover for survival.

The two British species of shad, the twaite shad and the allis shad, share an anadromous breeding cycle that has some similarities with that of Atlantic salmon in that they have an adult marine life stage but run a few clean and fast western British rivers to spawn. These shad species migrate up rivers in the spring to spawn in large shoals by night in fresh running water over sand or gravel bottoms. Spawned fish return to sea, fry progressively moving down rivers to reach estuaries during their first summer.

Perch spawn communally on hard submerged surfaces, depositing strings of sticky eggs. (Image credit: scubaluna/Shutterstock.com.)

The common sturgeon is also an anadromous species, though a scarce and generally solitary visitor to British waters in which spawning has not been recorded for many years. Female common sturgeon are 'broadcast' spawners, female fishes depositing 100,000 to three million small dark, sticky eggs at each spawning on sand, gravel and stones with no subsequent brood care. However, spawning does not occur every year as specific conditions are essential, including clear water over a shallow rock or gravel substrate and suitable water temperature and flow at the trigger point in spring determined by photoperiod (day length). After the eggs hatch, the larvae spend eight to fifteen days consuming their yolk sacs before emerging into the water column and being carried downstream into backwater areas. Here, the free-swimming fry spend their first year feeding on insect larvae and crustaceans before migrating into the main stem of the river and returning to estuaries and the open sea. Sturgeons are long-lived fishes with an average lifespan of 50 to 60 years, and some specimens living as long as one hundred years. They are also late maturing, first spawning not occurring until individuals are fifteen to twenty years old. These specific spawning requirements, late maturation, vulnerabilities over a long marine adult phase about which little is known and specific environmental requirements for development of juvenile fish contribute to significant conservation concerns about British and European populations of this now very rare fish.

## 7.10 REPRODUCTION AMONGST THE MARINE INTRUDERS

Those species entering the lower reaches of British rivers as transient visitors from the sea all spawn in saline water.

The three mullet species – thick-lipped grey mullet, thin-lipped grey mullet and golden grey mullet – breed at sea during the late summer or winter, releasing pelagic, non-adhesive eggs that float in the plankton. After hatching, juvenile mullet drift inshore, often inhabiting estuaries and coastal saltmarshes, to graze on the rich food that is found there in the summer months.

European seabass also spawn at sea, typically in early springtime through to June, at temperatures from 12°C–14°C, as day length is increasing. Bass produce small (approximately 1 mm), buoyant eggs often near estuaries or coastal areas. The eggs drift freely in the open water for around three days before hatching to produce small larvae measuring approximately 3 mm. Estuaries and salt marshes are favoured habitats for juvenile bass, shoals of 'school bass' feasting on invertebrates and small fish as they enter these habitats on a rising tide. Bass grow slowly, male fish maturing sexually after four to seven years and female fish doing so from five to eight years. This is a concern, as commercial exploitation of bass removes a substantial proportion of the stock before the fish have had a chance to breed.

Flounder also spawn in marine waters in the spring, but juvenile and adult fish routinely head inshore to feed in estuaries and coastal waters, including penetrating considerable distances upstream into fully fresh water.

The two entirely unrelated British smelt species, the smelt and the sand smelt, have a stronger association with estuaries when it comes to breeding. The smelt (*Osmerus eperlanus*) breeds in brackish waters, entering estuaries and occasionally lower rivers to spawn between February and April. Adult smelt return to sea after spawning, whilst the juveniles occupy the

Juvenile 'school bass' enter estuaries and salt marshes on a rising tide to feast on invertebrates and small fish.

estuary during early life stages, progressively moving down to the sea as they develop. The sand smelt (*Atherina presbyter*) spawns from May into July in coastal lagoons or large intertidal pools, some of which may be in the lower reaches of estuaries.

## 7.11  WEIRD SEX AND DEVELOPMENT

To conclude this chapter on *Sex lives of the British freshwater fishes*, we shall look at a few of the odder cases of reproduction and development (as if the bizarre life cycles of bitterling and European eels were not strange enough!).

### 7.11.1  Satellites in our rivers

Distributed right across the world, there are thirty-eight extant species of lamprey. Some, such as the brook lamprey found in British rivers, do not take to sea on metamorphosis nor do they feed, simply dying after spawning. Others, such as the river lamprey and sea lamprey metamorphose into marine parasites that return back to rivers to spawn before dying.

However, what is rather odd is that right across this global range, there are pairs of species, one sea-going and the other not, with ammocoete larvae that are, to all intents and purposes, identical. These are known as 'satellite species'. It is possible that, although they inhabit the same spawning and larval habitats, these different lampreys have been isolated from each other somehow as discrete breeding populations that are in the slow process of becoming genetically distinct (known as sympatric speciation). For example, it is known that lampreys choose mates of a similar body size, which may represent one method of segregation. But it is odd that this phenomenon occurs with different pairs of satellite species right around the world.

Ammocoete larvae of brook lampreys and river lampreys are indistinguishable. (Image © Dr Mark Everard.)

Emerging thinking based on DNA analysis seems to suggest that these pairs of satellites may represent different life strategies of one and the same species, much as sea trout and brown trout were formerly considered distinct species. Given the ancient evolutionary lineage of the lampreys, and that these ancient fishes have shown little change over 360 million years way back in the Carboniferous period, it seems unlikely that they are only just getting round to evolving! Many more mysteries of lamprey satellites need to be worked out, but currently emerging opinion is that pairs of satellite species may well be one and the same species with different life strategies.

### 7.11.2 Sex without 'real sex'

There is one odd freshwater fish that is a near neighbour, common across France and also much of continental Europe albeit not in British fresh waters, that has a mating strategy fascinating enough to deserve mention here. This is the gibel carp (*Carassius gibelio*), also known as the Prussian carp.

To produce viable young, gibel carp mate like other fishes do, in the springtime like most members of the carp and minnow family of which they are part. However, what distinguishes this ostensibly normal breeding behaviour is that the eggs develop without being fertilised. This is due to the phenomenon of 'gynogenesis'. Gynogenesis is distinct from 'parthenogenesis' as seen in aphids and some other insects, in which eggs hatch without mating explaining their rapid reproductive output. By contrast, mating is necessary for

**Gibel Carp found across continental Europe may be clones, produced by the phenomenon of gynogenesis. (Image © Dr Mark Everard.)**

gynogenesis. However, in this case, the sperm merely triggers embryo development without fusing with the cell nucleus of the egg. The developing embryo therefore contains only chromosomes from the maternal fish.

Odder still, the sperm triggering the development of eggs released by female gibel carp need not even be from other gibel carp. It seems that almost any similar fish species suffices. Genetic research has determined that gibel carp are triploid. (We humans and virtually all other non-microscopic living things are diploid, with pairs of chromosomes.) Triploidy can be triggered in the artificial breeding of rainbow trout and brown trout by heat shocking the eggs prior to fertilisation, producing sterile offspring that not only grow quicker in aquaculture but will not interbreed with wild stock when introduced into fisheries. However, gibel carp appear to be naturally triploid and so may possibly be infertile clones. This may explain to anglers why every gibel carp they catch in continental Europe looks almost identical.

Gibel carp are closely related to the crucian carp and also the goldfish. Fish breeders have found that it is possible to induce gynogenesis in goldfish, some of which end up as triploid fish. This may explain the evolutionary origins from crucian carp or goldfish of the hardy gibel carp.

Some species of catfishes, though not the wels catfish, are also known to practise gynogenesis. However, no instance of this phenomenon has thus far been discovered amongst British freshwater fish species.

### 7.11.3 Dressing to the right
The flounder is a member of the right-eyed flounder family (*Pleuronectidae*), so called because most species within this family lie on the sea bed on their left sides with both eyes on their right. When you look at a flounder, you are basically seeing a fish lying on its left side. On one 'side' of the body is a long dorsal fin, and on the other side a long anal fin, the mouth remaining sideways but both eyes migrating to the upper surface. However, flounders neither hatch nor live their early lives like that.

Flounder eggs float after fertilisation in the plankton of coastal waters, hatching out into juveniles no larger than a pinhead. Throughout their first few weeks of life, flounder fry look and act like other fish fry, with symmetrical bodies, swimming upright in the warm surface waters and feeding on small suspended animals. However, after just a few weeks of rapid growth, the body rotates and flattens as the young flounder head into deeper water. Progressively, the fish spend more time lying on their side, the dorsal and anal fins taking on their adult form and both eyes migrating to the upper surface. From this stage onwards, the flounder is wholly dedicated and adapted to life on the bed of the sea, often moving into estuaries and up rivers.

**Flounders start life as fry shaped like other fish, metamorphosing into their adult flatfish form with the eyes migrating to the right side of the body, though a proportion of flounders have their eyes on the left side of the body.**

However, life is not quite that simple (if such a drastic change can be described as 'simple'). A small number of species within the right-eyed flounder family can have their eyes on the left side, lying on their right side, notably members of the genus *Platichthys* which includes the flounder (*Platichthys flesus*). In fact, around 30% of flounders have their eyes on the left side, this proportion even higher in some populations.

# THE THINGS THAT BRITAIN'S FRESHWATER FISHES EAT

## 8

All fishes eat. However, it was not always assumed that this was the case. In *The Compleat Angler*, Izaak Walton's Piscator relates that,

> *Concerning which you are to take notice, that it is reported by good authors, that grasshoppers and some fish have no mouths, but are nourished and take breath by the porousness of their gills, man knows not how: and this may be believed, if we consider that when the raven hath hatched her eggs, she takes no further care, but leaves her young ones to the care of the God of nature, who is said, in the Psalms, 'to feed the young ravens that call upon him'.*

Some fishes were thought potentially able to absorb nutrition. But, of course, we now know that all fish eat to survive, as do all living things in one guise or another. However, a number of Britain's freshwater fish species do not eat as adults, or at least during certain phases of their adult lives.

It would be easy at this point to generalise about the diets of fish – filter-feeding, parasitic, predatory, herbivorous and omnivorous – but this would hide a lot of the complexity of the diets of Britain's freshwater fishes. With no exceptions, the diets of British freshwater fishes are more diverse than can be simply generalised. We will of course touch upon these five general feeding strategies, but also tell the story of how fishes really feed in 'warts and all' detail.

## 8.1 THE DIETS OF JUVENILE FRESHWATER FISHES

In our consideration of the life cycles of Britain's freshwater fishes as they become free-swimming, we have already observed their early dependence on essentially similar food sources, particularly small invertebrates. Many of the smaller cyprinids and other species may even start with food items that are smaller still, such as single algal cells.

Fish fry are amongst the smallest vertebrates found in nature and only partially developed once they become free-swimming. The gut of a dace fry, for example, is initially relatively undeveloped, comprising simply a straight tube from mouth to anus with no differentiated areas. The gut of a juvenile dace does not even develop a loop extending its effective length until about six weeks after hatching. The poor development of the digestive system in this tiny early life stage not only necessitates near-continuous access to prey, but the fry initially also depend on digestive enzymes released from the guts of their invertebrate prey when they are swallowed and masticated. This dependency on adequate and suitable small invertebrate food is common to all cyprinids and many other groups of fish.

As fish grow and develop, they move to progressively larger food items. The diet begins to diversify further as they become physically big enough to specialise in different food items. The diet of different fish species subsequently changes with their particular evolutionary adaptation, but also in response to food availability and the size of food item that the fish can ingest. As one example, roach grow slowly in the early years of their lives, attaining a length of just ten to fifteen centimetres

Daphnia and other types of 'water fleas' are important food items for the fry stage of most juvenile fishes and can be part of the mixed diet of some species into adulthood. (Image credit: Lebendkulturen.de/Shutterstock.com.)

after their third year of growth. Thereafter, the mouth of the roach develops a gape large enough to consume energy-rich molluscs, such as water snails and pea mussels that form a significant element of their adult diet, and from this point onwards their growth rate accelerates rapidly.

## 8.2 THE DIETS OF ADULT FRESHWATER FISHES

As freshwater fishes develop through larval and fry stages into progressively larger and more clearly differentiated juveniles and adults, they become progressively more dependent on the diets to which they are adapted. But, as ever, there are exceptions. In this section, we will look at five primary diets – filter-feeding, parasitic, predatory, herbivorous and omnivorous – before then considering where this food comes from, its timing and interactions with other elements of ecosystems.

### 8.2.1 Filter-feeding

Filtration of fine suspended microscopic organisms and other fine organic matter from surrounding water is a feeding strategy widespread amongst invertebrates. It is also the principal feeding strategy of the ammocoete larvae of Britain's three lamprey species, supplemented by grazing on detritus (as described later in this chapter). Whilst the river lamprey and the sea lamprey subsequently metamorphose to migrate downstream and live out their adult lives in estuarine and marine waters, the brook lamprey lacks mouthparts after metamorphosis and so is solely dependent to complete its life cycle on what it has fed on as a larva.

Video footage of shoals of shad at sea with mouths gaping reveals that they too filter-feed, using elongated gill rakers (extensions of gill arches oriented forwards inside the mouth cavity)

**Silver carp, a problematic invasive species in many countries though not established in Britain, filter-feeds on plankton as an adult. (Image credit: Sergey Goruppa/Shutterstock.com.)**

adapted for the task of sieving fine plankton. However, although both twaite shad and allis shad run some western British rivers to spawn in spring, it is not clear that they filter-feed in fresh waters. Some fall to anglers' flies and small lures when in fresh waters, suggesting that a degree of predatory feeding on larger food items does occur.

Whilst there are many species of filter-feeding freshwater fishes around the world, such as the silver carp native to China and Eastern Siberia but now widely distributed globally, no others occur in British waters. That said, there is a fine distinction between fry feeding on small animals and plant matter suspended in the water column, and large individuals filtering out these same small food items.

### 8.2.2 Parasitism

As observed above, river lamprey and sea lamprey metamorphose after several years as ammocoete larvae into adult, sea-going forms with a parasitic feeding strategy. The mouth of the adult form is cuplike, lacking jaws but instead armed with concentric rings of horny, keratinised teeth and a probing tongue. They attach themselves to the flanks of other fishes, rasping away tissues and secreting substances that prevent the victim's blood from clotting.

The lamprey then feeds on the blood (haematophagous feeding) and soft tissues of the parasitised fish, which can be any of a wide range of species and which then typically dies from excessive blood loss or infection. Some individual lampreys can even start haematophagous feeding in the river before migrating to the sea.

### 8.2.3 Predation

Many of Britain's freshwater fishes become predatory, to a greater or lesser extent. As observed above, the fry of all species depend initially on a diet rich in small invertebrate animals, even those species that we tend to think of as substantially herbivorous or omnivorous. For most species, invertebrates of varying sizes remain important constituents

Wound on the flank of a sea trout resulting from parasitic attack by a river lamprey.

of the diet throughout adult life. The common bream, for example, feeds primarily on small invertebrates such as chironomid (midge) larvae in the silt of slower-moving river and lake beds, their stable, laterally compressed body shape and fin arrangement enabling them to orient themselves to suck up these fine food items, separating them from the surrounding sediment with mouthparts that can be protruded like a tube. Tench have similar feeding habits, which are also deployed by many cyprinids and other fishes such as roach and dace.

Barbel possess thick, rubbery lips and two pairs of fleshy barbels (whiskers) around their downward-oriented mouths as an adaptation to sensing larger invertebrates in river sediments and under stones. They are also far from averse from eating any small fishes and plant matter that they locate.

'Bloodworms', the larval stage of midges, inhabit soft lake and river sediments and are favoured food items for common bream. (Image credit: schankz/Shutterstock.com.)

The mouth of a barbel is underslung on the snout and surrounded by two pairs of fleshy barbels (giving the fish its common name) adapted to root out food items on river beds using tactile and chemical senses.

Bleak, by contrast, tend to intercept fine food items in mid-water or at the surface, small aquatic and winged invertebrates forming a significant proportion of the diet. Large eyes and upward-pointed mouths adapt bleak ideally to exploit small animal food items in the spindrift. Rudd too are another cyprinid fish with an upward-oriented mouth, and they tend to depend significantly on picking out cladocerans (water fleas) and other small invertebrates from the water column. Cladocerans remain import food sources for rudd throughout their lives.

Brown trout and Atlantic salmon also live predatory lifestyles as juveniles and throughout adult life. The parr stage of these species can be territorial, though require specific riffle conditions (fast water over gravel), meaning that there is intense competition for space, limiting the number of individuals that any suitable patch of habitat can support (known as 'density

THE COMPLEX LIVES OF BRITISH FRESHWATER FISHES

The mouth of the bleak is upward-pointed and the eye is large, adapted to feeding from surface spindrift in rivers.

dependence'). Trout and salmon parr remain close to the river bed, rushing to intercept insects, small fish and other suitable small animal food washing down in the flow. As adults, brown trout and also Arctic charr maintain a predatory diet, brown trout continuing to feed on insects and other smaller fishes. Sea trout and salmon species smoltify and take to sea, living out their adult lives feeding on crustaceans such as krill, molluscs and also smaller fishes.

However, when most of us think of predators, we probably think first of species that feed on larger animals than invertebrates. A surprising number of Britain's freshwater fishes fit that bill.

Perhaps the best known and most infamous British freshwater fishy predator is the pike, sometimes dubbed the 'freshwater shark'. The configuration of the Pike is elegantly adapted to its lifestyle as an ambush predator. Its long, streamlined body with a cluster of dorsal, caudal and anal fins at the rear enables the pike to accelerate explosively, engulfing passing prey in its cavernous mouth armed with stout, sharp teeth in the jaws and vomerine teeth in the roof. As it lunges, the pike's cavernous mouth expands to 'inhale' the prey. Live or dead fish of any species, cannibalism being common, are predominant fare. However, pike will readily take ducklings and even fully grown ducks and other water birds, small mammals, amphibians and reptiles, as well as larger invertebrates. Pretty much any live or dead animal that will fit into that gaping mouth is fair game.

By contrast to the ambush tactics of the solitary pike, perch often shoal, particularly when smaller, and tend to be 'pursuit predators', running down their prey. It is not uncommon to see perch nipping at the tail of a fleeing prey fish, progressively disabling it before swallowing it whole. Though perch lack teeth in the jaws, the fine mesh of vomerine teeth in the roof of the mouth effectively latch onto prey fish, large invertebrates such as earthworms, amphibians and other prey items before they are swallowed.

**Perch tend to shoal and hunt small fish.**

Denizens deeper down in the dark depths include predatory zander and wels catfish adapted to hunting out prey fish in near or total darkness.

Zander, otherwise known as pike-perch after the two species they resemble, are predominantly piscivorous (fish-eating, at least as adults) members of the perch family. They are adapted to feeding on low light, enabling them to exploit food in murky waters, at depth or else nocturnally (though their peak activity is actually crepuscular, or in other words in the half-light of dusk and dawn). This is enabled by eyes with a reflective layer, the *tapetum lucidum*, behind the retina (light-sensing layer of cells), very much like that found in cats and other night-active animals as an adaptation to maximise vision in low light. In addition to the fine mesh of vomerine teeth on the roof of the mouth also found in other members of the perch family, including perch and ruffe, zander also have sharp teeth in the jaws including a prominent pair of 'vampire' teeth in the front of both the upper and lower jaws that they use to impale their prey.

Wels catfish by contrast have no teeth, but instead possess a cavernous mouth with jaws lined with horny plates to trap and crush prey animals – fish, water birds, mammals, large invertebrates – and all manner of other food items. The long 'whiskers' giving the catfish their common name are highly sensitive to chemicals in the water, as indeed is virtually the whole body and all of the fins of this large fish. The barbels are also highly tactile. Whilst the eyes of the wels catfish are minute, the slightest contact with those sensitive whiskers can trigger a lunge from the catfish to engulf would-be food items. In one South African river, I have regularly observed a similar but not closely related catfish species 'whiskering', or in other words swimming below the surface with just the tips of their whiskers visible as they sense the surface film, particularly at dusk, any contact with the finest potential food item inciting a violent 'take'. I have even caught 'whiskering' catfish when fishing with tiny artificial flies intended for other fish species. It is not certain if the wels

The eye of a zander looks 'dead', the reflective tapetum lucidum behind the retina maximising capture of light, enabling the fish to hunt in near darkness.

The barbels, or 'whiskers', of a wels catfish are highly tactile and chemically sensitive, helping this largely nocturnal species locate food in complete darkness.

catfish undertakes this form of whiskering behaviour, but sometimes they can literally put the 'cat amongst the pigeons'. These large predators, growing in excess of 100 kg (220 lb) and even larger (with a maximum published weight of 306.0 kg) in some continental European rivers, can outstrip food reserves and begin to hunt sometimes unexpected prey. One such example is a population of wels catfish in the Tarn River in southwest France, where they were introduced in the 1980s, that has started hunting pigeons that come to the river's edge to drink, sometimes nearly stranding themselves on the bank as they lunge at the thirsty birds. There is now a great deal of video footage of this remarkable and surprising behaviour on the internet, revealing catfish stalking their intended prey and engulfing, or dragging down by the foot, any pigeon that strays too close. There are also reports of large wels catfish attacking swimmers in some German lakes, though it is doubtful if these extremely rare encounters are an attempt at predation or rather a lunge at a flailing limb mistaken for smaller prey, or as a territorial reaction by nesting catfish.

The reality is that many British freshwater fishes are opportunistic predators and far from averse to eating fish or other animals when that food source is readily available. Many apparently non-predatory species will readily take fish fry when they are present in abundance, including, for example, roach, dace and common bream. Chub become notably predatory in warmer months, taking fish, small mammals and any other food items that will fit into their disproportionately large mouths. Barbel too can often be seen chasing minnows and other small fishes in the summer and will eagerly take ammocoete larvae and other small fishes located when probing soft sediment or under stones with their fleshy barbels.

Shocking at first sight are the many videos now on the internet of common carp opportunistically swallowing ducklings when both are feeding at the water's surface on offerings of bread as people feed ducks. A duckling fits easily into the mouth of a large common carp, and any duckling swallowed whole would be quickly squashed and ground by the powerful pharyngeal teeth in the throat of the carp, providing a substantial and protein-rich meal.

### 8.2.4 Herbivory

Although many freshwater fishes around the world are predominantly herbivorous – adapted to browse soft vegetation, scrape encrusting algae from stones and other hard submerged surfaces, filter algae from the water or ingest live plants – few British species adopt this habit. Whilst it is true that many species browse the submerged surfaces of shoots of emergent plants and underwater vegetation, rocks, wood and the sediment for food items significantly containing algae, they are in reality best considered as omnivores as small animals are as readily located and consumed there.

A notable exception is the grass carp, with a native range from northern Vietnam to the Amur River on the Siberia–China border but widely introduced globally including non-breeding stocked populations scattered across British fresh waters. The grass carp gains its name from its herbivorous habit and has been stocked around the world for biological control of aquatic weeds as well as in aquaculture for human food. That said, the grass carp is as much an opportunistic omnivore, including plants but not a dedicated herbivore.

Many other pieces of plant matter are consumed by various species of freshwater fishes on an opportunistic basis. Elderberries dropping into the water from surrounding trees are favoured by many species such as roach, dace and chub, as indeed are blackberries and other fruits

**Chub, like many coarse fish species, adapt their diet to available food, including, for example, the season glut of blackberries, elderberries, seeds and other plant matter dropping into rivers.**

which are exploited by chub though are rather too large to be accommodated by species with less cavernous mouths. Seeds falling or blown into the water are also a widely exploited rich food item, as are turions (overwintering buds that become detached from various aquatic plants to remain dormant until the following spring).

### 8.2.5 Omnivory

You could by now have formed the opinion from the preceding sections that the diet of many species is, in reality, plastic and attuned more to opportunism than a dedicated food source. Aside from dedicated parasites (adult river lamprey and sea lamprey) and a few obligate predators (pike, zander, perch, ruffe and bullheads, in particular), this is generally true.

For many species, diet is determined by what is available and limited by the gape of the mouth. Take, for example, the highly adaptable roach. As fry, fine algal cells and tiny invertebrates give way to progressively larger plant and animal food, the growth rate accelerating markedly when the mouth gape of the growing fish is wide enough to access a richer larder of larger molluscs and other invertebrates. And, of course, adult roach are far from averse to making a meal of abundant fry shoals in river and lake margins in warm weather, potentially consuming some fry of their own species. Of all of Britain's freshwater fishes, the chub is perhaps the most notoriously omnivorous, a robust cyprinid with a disproportionately large, toothless mouth oriented forwards on the head. Chub seem happy to engulf just about any plant or animal matter from detritus to soft plants, worms, small fish and mammals or pretty much any available food item from the bed, mid-water or surface.

Detritus is a surprisingly common food for many freshwater fishes. Detritus comprises generally brown, amorphous organic matter common in fresh waters, resulting from breakdown of plant and animal matter. Ammocoete larvae of lamprey species can, in addition to filter-feeding, browse detritus from the surface of the soft sediments in which they lie buried. The three British species of mullet that penetrate lower rivers are also largely detritivores (species that consume detritus), seen moving up with the tide or residing in lower rivers in warmer months where they browse on layers of detritus on sediment and hard surfaces.

During winter months, when other food is scarce and many water plants have died down, the rich growth of microscopic fungi, bacteria, protozoa and other microorganisms growing on and within the detritus is a nutritious food source for many fish species, particularly the cyprinids.

### 8.3 WHERE DOES ALL THIS FOOD COME FROM?

So, from where does this rich larder arise? Clearly, piscivores and parasites feed on fish produced from within their immediate environment or migrating into it. However, the source of food is far less clear-cut for filter-feeders, predators, herbivores and omnivores.

**Nutrient-poor upland streams tend to be allochthonous, much of the food entering the water body from the surrounding landscape. (Image © Dr Mark Everard.)**

As a very general rule, purer, more turbulent headwaters of rivers and the small pools located in nutrient-poor upland areas are not highly productive. Consequently, the quantity of potential plant and animal food produced within the water body itself is sparse. Yet, these waters can hold an abundance of fish, particularly juvenile salmonid species hatched from well-flushed spawning redds. In these situations, much of what the fish eat is allochthonous, or in other words arises outside the water body. Seeds, flying and wind-blown insects, worms and other invertebrates flushed into the water make a ready meal. In these low-nutrient systems, many of the aquatic invertebrates that can form important food items also feed on organic matter washed into the system or predate on other invertebrates in the stream.

By contrast, mature lakes and slow, lowland rivers tend to be highly productive. These are known as autochthonous systems, in which an abundance of plants and a diversity of invertebrate and other animal food items is produced within the water body.

Of course, life is never so cut-and-dried. Bankside trees surrounding lowland pools and rivers attract fish not merely for refuge, spawning sites and food on the submerged roots, but also as a rich source of insects and other potential food items falling into the water from the overhead leaves and twigs. Likewise, amphibians, reptiles, small mammals, worms, ducklings and other fare grown fat on dry land may be eagerly engulfed by lurking predatory and opportunistic fishes.

Extreme weather, too, opens alternative larders of benefit to fishes. In drought conditions, retreating water levels may strand freshwater mussels and other invertebrates in river and still water sediments, providing food to terrestrial animals but also for fishes and other river life when water levels rise again. Spates too might turn rivers into boiling torrents and may make

Well-vegetated lowland rivers tend to be productive, autochthonous systems, much of the food deriving from within the river system. (Image © Dr Mark Everard.)

still waters murky, but they also wash in all manner of invertebrates, particularly including slugs and worms when banks collapse, as well as a wealth of seeds and other potential food.

Flowing and standing waters are local manifestations of a greater water cycle operating across whole landscape and atmospheric systems. Freshwater fishes, including their diets, are adapted to opportunistic exploitation of the bounty of integrated land/water catchment systems, wherever within them the available food items are produced.

## 8.4 TIMING

The study of the timing of plant and animal life cycle events is known as phenology. Well-known, long-term phenological records – first springtime emergence of leaves and flowers, butterflies, the call of the cuckoo and the reappearance of swallows and martins – highlight two things. Firstly, there is a natural rhythm and synchrony linked to the seasons. Secondly, we know that the timing of these events is shifting, providing further evidence of the changing climate. From phenological observations, amongst an overwhelming body of corroborating evidence, we know that we are living in a time of profound climate change.

This matters because the myriad interconnections within nature are elaborately co-evolved. The seeds and berries of plants feed birds and mammals in preparation for overwintering, and emerging aquatic insects in springtime feed juvenile fish and post-hibernation bats and arriving summer-migrant birds. Flooding cycles enable spawning fish to access marginal wetlands, the nectar of flowers feeds insects of the same season to mutual advantage and the nourishment of linked food chains and so on. Break the timings and the functions of nature, for all its evolved adaptive capacities, inevitably break down.

As we have seen, adequate and suitable fine invertebrate food is vital for the success of most early stage fish fry. The spawning period of different freshwater fish species is adapted for the larvae to exploit available resources or to evade predation. Salmonid species mainly spawn in winter in cool gravel redds, the fry 'swimming up' in synchrony with the spring abundance. Most other coarse fish species spawn in warming waters during spring and early summer, again to exploit this glut. With the exception of burbot, winter-spawning but now extinct in Britain, the earliest spring-spawning species are pike and dace, both often as early as late February or March. Dace grow well in cool water and steal a march on access to food over competitor species. Pike spawn early for another reason: the fry are strongly piscivorous from birth and ready to gorge on the fry and juveniles of freshwater fish species as they emerge later in the year.

If the fine synchronisations evolved over millions of years become disconnected, prospects for the survival, breeding success and resilience of fish populations and ecosystems to environmental extremes and stresses can be seriously compromised. Evolution has equipped

Juvenile pike hatch from eggs laid early in the year, predating on the young of other fish hatching later in the year as well as other pike.

fishes with strategies to deal with sporadic events, such as unseasonal floods that wash out fry and their food alike in any one year, but sustained change has major adverse implications.

## 8.5 DIET AND ECOLOGY

It is through feeding and being fed upon that organisms most profoundly interact with their environment and play integral roles within the ecosystems of which they are part. Matter and energy are thus cycled, made available further along complex food webs, and populations of species are regulated through long-evolved checks and balances. That fish are often a neglected element of aquatic habitats is to the detriment of wise ecosystem management.

Some nature reserves, including, for example, certain lakes on the Cotswold Water Park in Gloucestershire, are maintained to ensure adequate provision of small fish to support rare visiting birds. A particular conservation target here is the piscivorous bittern (*Botaurus stellaris*), a member of the heron family formerly common in Britain and west and central Europe up until the nineteenth century when many breeding areas were abandoned because of drainage and persecution. Reserve managers are less aware of the species of fish that reed and water management favours (in fact they are mainly rudd), the purpose of measures to promote the fish being purely to feed bitterns. How much greater ecological diversity and resilience could be achieved were mixed fish populations managed consciously, rather than simply regarded as fodder?

There are many interesting interactions between fishes and other elements of wildlife. One such is seen in the science and practice of biomanipulation. Simply put, biomanipulation is

The bittern is a scarce bird of wetlands that feeds on small fishes; some nature reserves managed to provide fish for them to feed on in their favoured reed bed edge habitat.

concerned with manipulating biological communities to change the characteristics of whole ecosystems. In some interesting cases, mainly experimental in the Britain though applied practically in some of the Norfolk Broads and large still waters elsewhere in Europe, fish populations have been managed as part of a wider programme to recreate former clear-water lake conditions with a healthy growth of aquatic plants, providing habitats for diverse wildlife. Removal of plankton-eating fish species such as roach, rudd and common bream has been undertaken in these locations, supported by measures such as suction-dredging of nutrient-rich mud out of the Broad beds and importantly also reducing further concentrated nutrient inputs from sewage treatment works and poor agricultural land use practices. The purpose for which these actions are undertaken is that these fish species feed on the 'water fleas' and other zooplankton (small animals in the plankton) that graze on the small planktonic algae responsible for opaque water conditions. As the zooplankton population controls algal density, a booming population of water fleas has the effect of clarifying the water. Greater light penetration can promote the growth of rooted water plants, encouraged by removal of fishes such as common carp and common bream that grub up sediment when feeding. Recovering stands of submerged water plants then absorb nutrients from the water, limiting planktonic algal growth and also harbouring sight-feeding predators such as perch and pike that then control the numbers of smaller species of fish. In this ideal situation, the former clear-water status of these water bodies can be restored together with the many ecological and socioeconomic benefits that cleaner water brings.

Regrettably, a more common occurrence is the introduction of species such as common carp, which rapidly convert formerly clear, well-vegetated and biodiverse still waters into turbid conditions through their voracious feeding habits. These habits include grubbing up

**Clear-water, rooted water plant-dominated lake ecosystems favour diverse fish and other wildlife communities, with water also less expensive to treat for human uses.**

the sediment, uprooting water plants and resuspending silt into the water column, effectively 'flipping' the whole ecosystem into a clouded-water state. Aside from the direct and continuing activities of the large fish perturbing the sediment, clouded waters in turn favour smaller fishes freed to some extent from the controls of sight-feeding predators, so reinforcing the change in ecological regime. Once 'flipped' from clear to cloudy, less biodiverse and economically more costly to clean water conditions, restoration becomes a very costly and contentious challenge.

Another of the many interactions between the diet of fish and that of other components of ecosystems is seen in spring after ducklings hatch. Watching ducklings dabbling in the spring, you may see them suddenly speed across the water when they spot a fly or midge, chasing it as it emerges from the water's surface or leaping to snatch it from the air with great precision. Ducks such as mallards (*Anas platyrhychos*) are generally thought of as herbivorous. However, in reality, they have an opportunistic, omnivorous diet that can include vegetation, seeds, acorns and berries, as well as some insects and shellfish. To support their rapid growth, protein-rich midges and other water flies form an important element of the early diet of ducklings and, if absent, can lead to malnourishment and poor growth, potentially including direct mortality but more likely greater vulnerability to their many would-be predators and diseases. Where dense populations of insectivorous fishes, such as common bream that feed extensively on 'bloodworms' (red-pigmented chironomid midge larvae in the silt of river and pool beds), significantly limit the quantity of flies emerging from the water, growth potential and long-term survival of ducklings can be compromised. This is yet another rationale for thinking in terms of the balance of the whole ecosystem, including their mixed fish populations, in management measures.

# THE CURIOUS
# WORLD OF
# BRITAIN'S
# FRESHWATER
# FISHES

The world of freshwater fishes is full of extremes, oddities, fables and curiosities. All the more reason, then, to take better account of this much-neglected component of our island wildlife.

## 9.1 THE BRITISH FRESHWATER FISH 'BOOK OF RECORDS'

Some interesting facts emerge when we think of the British freshwater fish 'Book of Records', or in other words some of the extremes of size, distribution and other facets of our fascinating freshwater fishy fauna.

### 9.1.1 Britain's biggest freshwater fish

Across the world, there are many competing claims for the title of the largest freshwater fish. Candidates include the Mekong giant catfish (*Pangasianodon gigas*), also known as the pla buk (literally 'huge fish'), naturally occurring in the enormous Mekong River Basin and with a record reported weight of 646 pounds (293 kg) and length of 10 feet (3 m). The Mekong is also home to an impressive range of other freshwater giants, including giant carp (*Catlocarpio siamensis*) that can weigh up to 660 pounds (300 kg). In South America, the arapaima (*Arapaima gigas*), also known as the pirarucu, can weigh up to 485 pounds (220 kg) and reach a length of 14.8 feet (4.5 m). In China, the giant Chinese paddlefish (*Psephurus gladius*), also known as the Chinese swordfish or 'elephant fish' due to the resemblance of its snout to the trunk of an elephant and even the 'giant panda of the rivers' due more to its rarity and protected status, has been reported as in excess of 7 m (23 feet) long and 1,100 pounds (500 kg). However, the mighty Chinese paddlefish appears to be already extinct due to compound human pressures. Another paddlefish, the American paddlefish (*Polyodon spathula*), found in slow-flowing waters of various major rivers of the US, with a characteristic flat bill extending out in front of the fish's large mouth that gives the fish a variety of local names, such as 'spoonbills', 'spoonies' or 'spoonbill catfish', can commonly exceed 1.5 m (5 feet) in length and weigh in excess of 60 pounds (27 kg) but with an official record of 144 pounds (65 kg). The sturgeons comprise another group of often large fish, Europe's common sturgeon (*Acipenser sturio*) said to have been taken to weights of 460 lbs (209 kg). Other of the world's sturgeon species grow even bigger, with documented catches of specimens (of species unspecified) in excess of 1,000 lb (454 kg), as well as tales of catches over 2,000 lb (907 kg) seen in some old-time pictures. Another Mekong River leviathan is the giant stingray (*Himantura chaophraya*), reported (though unsubstantiated) as reaching 1,100 pounds (500 kg) in weight and 16.4 feet (5 m) in length. Other massive freshwater stingrays are found in the Amazon River.

There is no clear winner here in substantiated claim to the crown of 'biggest freshwater fish', but sadly all share the same legacy of declining or extirpated stocks due to a combination of their great size, age at first spawning and vulnerability to overfishing, blocked migration routes and wider environmental degradation. These fishes, however, are mere minnows in the face of the largest fish known to science, fossilised bones discovered by palaeontologists in clay

**Sturgeon of prodigious proportions were more common in historical times, such as this common sturgeon from the Hundred Foot River in Cambridgeshire caught in 1906.**

pits in 2003 near Peterborough, England, substantiating a fish identified as *Leedsichthys problematicus* that swam the world's oceans some 155 million years ago in the Jurassic period with an estimated intact length of 22 m (72 feet).

By contrast, the biggest freshwater fishes found today in British fresh waters are more modest, but no less impressive set in the geographical context of our small islands. The biggest rod-caught Atlantic salmon of 64 lbs (29 kg) was landed on 7 October 1922 by a lady, Miss Georgina Ballantine, from the River Tay near Caputh in Perthshire. A 50 lb Atlantic salmon was caught by netsmen on the Exe estuary, now celebrated on a wall mural in the Devon town of Topsham. As a rod-caught fish, Miss Ballantine's Tay salmon remained uncontested as Britain's largest freshwater fish until September 2007 when a 69 lb 8 oz wels catfish was landed, again by a female angler. However, this giant catfish was not ratified as a British rod-caught record as the record list for the species had by then been suspended because of a thriving trade, both legal and illegal, in imported catfish in excess of the prior British rod-caught record.

Nevertheless, these mighty Atlantic salmon and wels catfish are dwarfed by the capture on 25 September 1933 of a common sturgeon of 388 lb (176 kg) from the River Towey in Camarthenshire (Wales) by a Mr A. L. Allen, who hooked it inadvertently whilst fishing for sea

Miss Georgina Ballantine poses with Britain's biggest rod-caught Atlantic Salmon of 64 lbs, landed on 7 October 1922 from the River Tay near Caputh in Perthshire.

trout. After a mighty battle, the sturgeon of fully 9 ft 2 in (2.79 m) with a girth of 59 inches (1.5 m) was landed. It remains the biggest fish caught from British fresh waters to this day. It is not impossible that a larger common sturgeon might fall inadvertently to rod and line angling, though this is increasingly unlikely given the elusive and increasingly threatened status of this royal fish.

### 9.1.2 Britain's smallest freshwater fish

Freshwater fish fry are amongst the smallest of all vertebrate animals. As we have seen, they hatch in a semi-embryonic state, still connected as larvae to their yolk sac, remaining with largely undeveloped fins, guts and bones in their early free-swimming days. The smallest British freshwater fish species when fully grown is a closely competed contest, with many 'small fry' species attaining just a few centimetres body length at maturity. The shy and elusive spined loach is often considered the smallest, yet this elongated and strongly laterally compressed fish can reach a body length of as much as 14 cm. Body length of the two British freshwater species of stickleback are shorter than this when fully grown. The three-spined stickleback attains a body length of up to 11 cm, but the ten-spined Stickleback tips the 'smallest British freshwater fish' record at just 9 cm long. There is, in fact, a tie for this mini-record attainment with the 9 cm-long sunbleak, though this is a non-native and problematic invasive species so, really, it should not count.

The ten-spined stickleback appears to be Britain's smallest native freshwater fish.

It is worth noting that our smallest species are giants compared to the world's smallest freshwater fish. This fish is *Paedocypris progenetica*, a tiny cyprinid that was only discovered in around 2006 in the peat swamp forests of Sumatra. It measures just 7.9 mm from nose to tail. This is less than one-tenth the length of the ten-spined stickleback and around the size of a large mosquito larva. *Paedocypris progenetica* is also the world's smallest vertebrate. Not only that, this remarkable fish has evolved to live in waters stained dark with decaying tropical vegetation and with an exceptionally high acidity with a pH as low as 3 (roughly that of vinegar). It has a see-through appearance with a reduced bone structure around the head that leaves the brain unprotected by skeleton. A very remarkable freshwater fish indeed, albeit one now at considerable threat from habitat loss.

### 9.1.3 Britain's most northerly freshwater fishes

Though freshwater fishes are found in other countries far closer to the North Pole than those in Britain, two species found in Britain may share the claim to be the world's most northerly freshwater fish. The three-spined stickleback is particularly hardy, adapted, for example, to take to sea during winters when rivers rise into intolerable spates or else freeze. They are present at many high northerly latitudes. However, exceeding even the tough three-spined stickleback is the Arctic charr. Deriving its name from the habitats to which it is adapted, the Arctic charr is the most common and widespread salmonid fish in Iceland and is the only species of fish in the far-northern Lake Hazen, the northernmost lake in Canada on Ellesmere Island in the Canadian Arctic.

### 9.1.4 Britain's deepest freshwater fishes

Another extreme in which fish occur is in deep lakes. The deepest lake in the British Isles is Loch Morar, on the west coast of Scotland, with a maximum depth of 309 m (1,014 feet). Loch

The three-spined stickleback may be Britain's most northerly native freshwater fish.

Morar is considerably deeper than Loch Ness, the second-deepest lake with a maximum of 230 m (754 feet). South of the border, England's deepest lake is Wastwater to the southwest of the Cumbrian Lake District, a rift lake (formed by a split in the Earth's crust) descending below sea level to a maximum depth of 76 m (249 feet). Both Loch Morar and Wastwater are home to Arctic charr as well as good stocks of brown trout. Loch Morar also receives a run of Atlantic salmon and sea trout.

### 9.1.5 Britain's most widespread freshwater fishes

Many species of British freshwater fishes are generalists, able to survive in a wider range of flow regimes and salinities and to adapt to diverse diets depending on available food. My 2006 book *The Complete Book of the Roach* celebrates this hardy and adaptable cyprinid species as 'the supreme generalist', as at home in a duck pond as a mighty salmon river, and thereby becoming the most widely distributed of the larger species of fish found in British fresh waters.

The Arctic charr is found at the greatest depths in British fresh waters.

However, an unhelpful and frankly patronising distinction is often made between these larger and more conspicuous species and the 'minor species'. And it is to one of these so-called 'minor species' to which the crown of Britain's most widespread freshwater fish is owed. The species in question is the three-spined stickleback, equally at home in diverse types of habitat from ditches and muddy pools, canals and lakes, reservoirs and river margins, tolerant of a significant degree of pollution and also of varying salinities even into full salt water in estuaries and coasts.

The extent to which humanity has messed with natural ecosystems, and their consequent potential consequences for supporting human wellbeing, is indicated by the now widespread introduction and naturalisation of brown trout from their native northern paleo-arctic range. Brown trout now occur in places as diverse as across the Himalayas, India's Western Ghats and other uplands of Africa, New Zealand and Australia, North and South America, the Falkland Islands, the Middle East including Turkey and the Jordan River, Japan, Pakistan, Swaziland, Fiji, Papua New Guinea and Cyprus. So too, rainbow trout have become globally widespread at the hand of humanity.

But perhaps the most globally pervasive of species, introduced as 'pigs with fins' in aquaculture and increasingly for sport, are three carp species. Two of these occur as introductions in British fresh waters – the common carp and the grass carp – whilst, mercifully, the silver carp, another globally problematic invasive species, is not naturalised in British fresh waters.

### 9.1.6 Britain's rarest freshwater fishes

To be absolutely strict about it, there are only really two candidates to claim the 'record' of the rarest British freshwater fish species: houting, declared globally extinct in 1982, and burbot, also extinct in British waters since the 1950s or 1960s. However, let us turn our attention to extant species.

Of the species listed on the IUCN (International Union for Conservation of Nature) 'Red List' of threatened species, we find three British species listed as Critically Endangered (CR). These include the European eel based on recent rapid population decline, the common sturgeon that in reality is anything but common and the gwyniad, recognised on the Red List as a separate species though more generally considered a local Welsh population of the European whitefish. We have to remember that this classification assesses species across their global range rather than distinctly for Britain. On the Red List, the vendace is listed as a species of Least Concern (LC), yet as we have seen, the English population occurred in only two lakes in the English Lake District with the Bassenthwaite population concluded as extinct in 2008. In Scotland, the two natural populations were declared extinct, with only a stocked population in Loch Skeen now established as a conservation measure. By any measure, the vendace is far rarer in British waters than its European whitefish counterpart and the European eel, though the parlous status of none is cause for complacency.

One fish that is commonly considered as Britain's rarest fish based on its localised distribution, lack of mobility, elusive habits and vulnerability to over-management of vegetated drains and backwaters is another 'minor species': the spined loach. Though also listed as a species of Least Concern (LC) by the IUCN Red List based on its global distribution, the spined loach is highly restricted in distribution in Britain, occurring in just eastern-flowing river catchments (the Welland, Nene, Great Ouse, Witham and Trent) in eastern counties of England. Furthermore, across this limited range, scientific analysis has revealed small genetic differences between fishes found, on the one hand, in the Welland, Nene and Great Ouse systems and, on the other, those from the Rivers Witham and Trent. This is thought to be related to the low capacity for dispersal of these small fish that can swim only weakly and tend to burrow or remain otherwise inactive.

The spined loach is commonly considered to be Britain's rarest extant freshwater fish.

It is hard to call an outright winner in the 'rarest species' category of records, different scarce species differing widely in characteristics and vulnerabilities. Perhaps the most important thing to take from this is that the requirements of these species need more adequately to be taken into account in the ways in which we use and manage the natural world, such that their genetic legacy and roles in ecosystems remain into the future.

### 9.1.7  Britain's oldest freshwater fish

An October 2003 newspaper article announced the sad demise of Goldie, said to be the oldest pet goldfish in the world. Goldie, exceptionally, reached a ripe old age of 45 years, the last survivor of three goldfish won in a Devon fair in 1960. This family pet was found resting on the bottom of the tank in Devon and was buried in the owners' garden. Even beyond death, Goldie brought pleasure to people as, following an unexpected upwelling of public sympathy, the bereaved owners asked for donations to sponsor an aquarium in an independent day school in Exeter for children with significant physical difficulties.

Goldie, of course, was a captive fish. In the wild, there are talks of common carp reaching venerable old ages, though the oldest recorded age in the scientific literature is a more modest, yet still admirable, 38 years. The oldest recorded age of a wels catfish is a lot longer at 80 years. However, it is doubtful that this non-native species, at the extreme north of its tolerance range where introduced in British waters, reaches anywhere near this age.

A pet goldfish set a record as Britain's oldest fish on its death at 45 years old in 2003, resulting in unexpected upwelling of public sympathy. (Image credit: Andrey Armyagov/Shutterstock.com.)

A koi carp in Japan died at the age of 226 in 1977, having lived in the same pool and with generations of the same family throughout its whole long life. The koi was already 25 years old when America made public the Declaration of Independence in 1776, survived empires as they rose and fell, and lived through two world wars.

By contrast, the oldest known freshwater fish was a koi carp (an ornamental variety of the common carp) known as 'Hanako', which died at the age of 226 in July 1977. Hanako had been handed down through generations and, amazingly, was already 25 years old when America made public the Declaration of Independence in 1776. Whilst empires rose and fell across the world, and technology leapt ahead at astonishing pace, Hanako had lived out its entire placid life in the same pool by the same house with the same family.

Another noteworthy fish of a species native to Britain that lived an abnormally long life is an eel that, according to a report in *The Telegraph* on 6 November 2017, died at the ripe old age of 155 after living in a family well in the little Swedish fishing town of Brantevik. This eel, thought to be the world's oldest, was found dead in the family well where it had lived since Napoleon III was King of France and work had just started on the Suez Canal, enduring two world wars. Before public water systems were developed in the 1960s, it was common practice to drop eels into household wells to get rid of flies and bugs. This record eel had been more of a family pet in its dotage and was reported to be survived by another eel living in the well that was believed to be a mere 110 years old.

### 9.1.8 Britain's hardiest freshwater fishes
Various freshwater fish species require high oxygen concentrations to thrive, including grayling, Atlantic salmon, and Arctic charr, as well as species of whitefish being particularly susceptible to declining oxygen concentrations in water and too long exposure out of water. However, some of Britain's freshwater fish species have evolved to withstand prolonged periods of depressed oxygen.

The spined loach can gulp a bubble of air to supplement oxygen absorbed by its gills. However, Britain lacks any freshwater fish species capable of surviving entirely by breathing air, though

A European eel died in 2017 in Sweden at the age of 155, thought to have been the world's oldest eel, having lived in a family's well since Napoleon III was King of France.

many do globally, enabling them to inhabit highly enriched, oxygen-poor waters. Such fish include the arapaima, native to the Amazon, the African bonytongue found across northern Africa, various lungfish species from Africa and Australia, many species from the Anabantidae family, including the Siamese fighting fish (*Betta splendens*), and various gourami species familiar as tropical aquarium species. The electric eel from South America also gulps air, and the Canterbury mudfish (*Neochanna burrowsius*) of New Zealand has a permeable skin enabling it to absorb around 40% of its oxygen needs across the body wall.

Tolerance of prolonged low oxygen levels is clearly important for fishes that live in waters that may freeze over seasonally. Tench can sometimes be found embedded in mud where dredging activities occur on lowland drains, the species thought to become torpid throughout the cold months. In *The Compleat Angler*, Izaak Walton relates that,

> *Gesner reports, that in Poland a certain and a great number of large breams were put into a pond, which in the next following winter were frozen up into one entire ice, and not one drop of water remaining, nor one of these fish to be found, though they were diligently searched for; and yet the next spring, when the ice was thawed, and the weather warm, and fresh water got into the pond, he affirms they all appeared again. This Gesner affirms; and I quote my author, because it seems almost as incredible as the resurrection to an atheist: but it may win something, in point of believing it, to him that considers the breeding or renovation of the silk-worm, and of many insects. And that is considerable, which Sir Francis Bacon observes in his History of Life and Death, fol. 20, that there be some herbs that die and spring every year, and some endure longer.*

Tolerance of low oxygen conditions also enables hardy species such as common bream and tench to thrive in nutrient-rich waters in which dense algal blooms may supersaturate the water on a bright day with as much as 180% oxygen due to their photosynthetic processes, but which respire at night to depress oxygen levels to 20% or less on a warm summer night.

The European Eel is also a hardy fish, able to remain out of water for some time, and even to make its way across land on a wet night when undertaking migrations or accessing isolated pools. The common carp is also noted for its hardiness, and its capacity to survive for some time out of water if kept moist. This may account for reports of these carp being transported long distances by monks, including into Britain as a cultivated food source.

However, the hardy crucian carp surpasses this durability, adapted to living for months in completely oxygen-free conditions. Across Scandinavia, where pools can freeze over for half the year with decaying vegetation and other organic matter progressively depressing oxygen concentrations in the absence of contact with the air, crucian carp thrive in the absence of competition with any other fish species. Extraordinarily, crucian carp have been found to have gills that change in structure according to ambient oxygen concentration. In oxygenated conditions, the lamellae (small, blood-rich plates on the gills that are the primary site of oxygen uptake) are embedded in the cell mass of the gills, presenting only as small respiratory surface area. But when the fish are transferred to or kept in oxygen-poor water, a significant reduction in the cell mass of the gills exposes the lamellae, substantially increasing the respiratory surface area. This change is reversible, enabling crucian carp better to cope with low-oxygen conditions in which other fish species are unable to compete or survive. The blood of the crucian carp also has a much higher affinity for oxygen than any other vertebrate. Even more remarkably, as oxygen supplies become even more limited, the cellular biochemistry of the crucian carp enables the fish to metabolise sugars without the use of oxygen, changing chemical pathways to excrete alcohol through the gills in order to avoid the build-up of dangerous levels of lactic acid. This

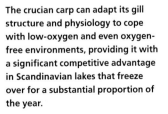

The crucian carp can adapt its gill structure and physiology to cope with low-oxygen and even oxygen-free environments, providing it with a significant competitive advantage in Scandinavian lakes that freeze over for a substantial proportion of the year.

truly astonishing combination of mechanisms enables crucian carp not merely to survive, but to maintain physical activity, even for a period of several months depending on temperature.

## 9.2 ROYAL FISHES

A number of Britain's freshwater fish species have certain royal or other ruling class connections.

### 9.2.1 The royal sturgeon

*Commentaries on the Laws of England*, an influential eighteenth-century treatise on the common law of England written by Sir William Blackstone, includes a passage outlining that the '*superior excellence*' of whale and sturgeon made them uniquely suited for the use of the monarch. Dolphins and porpoises have also taken on the common law designation 'royal fish', under which all of these 'fish' species (obviously whales, porpoises and dolphins are mammals) are the property of the monarch, either when caught or when the catch arrives on British shores from any other location. The right of the British monarch to 'royal fishes' was formalised by a statute put in place during the reign of Edward II (acceding to the throne in 1307 and deposed in 1327).

At the time that an independent Scottish government was established in July 1999, it became responsible on behalf of the Crown for dealing with royal fish stranded on Scottish shores, a judgement deeming that this applies to stranded whales measuring more than 25 feet long from the snout to the middle of the tail. The legal status of royal fish continues to provoke periodic controversy and interesting case law.

Across the English Channel, kings of Denmark and dukes of Normandy enjoyed similar privileges relaying to 'royal fishes'.

The common sturgeon is one of a number of 'royal fish' in Britain and elsewhere in Europe.

### 9.2.2 Imperial pets

Though not relating directly to Britain, it is interesting to relate an entry in Izaak Walton's *The Compleat Angler* that Sir Francis Bacon,

> ...*in his History of Life and Death, mentions a Lamprey, belonging to the Roman emperor, to be made tame, and so kept for almost threescore years; and that such useful and pleasant observations were made of this Lamprey, that Crassus the orator, who kept her, lamented her death; and we read in Doctor Hakewill, that Hortensius was seen to weep at the death of a Lamprey that he had kept long, and loved exceedingly.*

It seems that long-lived imperial fishy pets are the stuff of great affection and legend. However, as we will see later in this book, this may be a case of mistaken identity, as a similar word covered both eels and lampreys. It is difficult to see how a lamprey could be seen as a good pet given the need to keep giving it live (shortly to be dead) fish to satisfy its parasitic feeding habit

### 9.2.3 A dish favoured by royalty

Lampreys are considered a delicacy in some parts of Europe, including southwestern France, though are not commonly eaten in the Americas. Lampreys, as we have seen, are not a 'true' fish as they lack jaws, bones and paired fins.

Regardless of their ichthyological status, lampreys were nevertheless favoured by monarchs in English history. King John (1166–1216), who was King of England from 1199 until his death, was particularly fond of lamprey pie. Indeed, he is said to have fined the City of Gloucester the equivalent of £250,000 for failing to deliver his Christmas lamprey pie. Various accounts suggest that King John finally died of 'a surfeit of peaches and cider' shortly before his fiftieth birthday.

**Lampreys are considered a delicacy in some parts of Europe.**

### 9.2.4 Death by 'a surfeit of lampreys'

Another devotee of lampreys was King Henry I (1068 or 1069–1135), the fourth and youngest son of William the Conqueror and King of England from 1100 until his death. Travelling to Lyons-la-Forêt in November 1135 to enjoy some hunting, King Henry fell ill and, the chronicler Henry of Huntingdon reports, ate a number of lampreys against his physician's advice, leading to the worsening of his condition over the course of a week until his death.

In medieval thought, items of food were assigned one of four humours with an associated moisture and temperature: sanguine (warm and moist), yellow bile (warm and dry), phlegm (cold and moist) and melancholic (cold and dry). Akin to contemporary yin-yang balances in China and Ayurvedic practices in India, diet and wider lifestyle had to be kept in balance, with an equal mix of all elements. The physician's advice would have accounted for the imbalance of lampreys with the humour already perceived as already in excess in the sickening king. The chronicler Roger of Wendover (died 1236) reported that the king '...*would not listen to their advice. This food mortally chilled the old man's blood and caused a sudden and violent illness against which nature struggled and brought on an acute fever (trying to warm the body up) in an effort to resist the worst effects of the disease*'. Perhaps the humours were out of balance, the fish in poor condition causing food poisoning or the meal was coincidental with another underlying cause of death.

'A surfeit of lampreys', a dish for which King Henry was known to be excessively fond, may be unique in leading to a monarch's demise.

### 9.3 ICONIC FISHES

The scientific term 'iconic species' is one that I coined when considering river conservation in a publication with my friend and colleague Gaurav Kataria in India. Many of us will be familiar with the term 'flagship species' in nature conservation, for example, with a focus on pandas, tigers, elephants and other species with complex needs and connected habitats serving as a 'flagship' for conservation action not only for these habitats and their connectivity but for the many species these habitats support.

However, 'flagship' was an insufficient term for fishes, not the least reason being that they are by and large also directly exploited for food, as cultural icons and indicators of the health of water bodies and a range of other reasons. 'Iconic' was a more appropriate term and is one used in this section to address both its conservation meaning and its more common cultural association. A number of freshwater fish species occurring in Britain have an iconic place in both nature conservation as well as festivals and celebrations.

'The Salmon is accounted the King of freshwater fish'.

### 9.3.1  King of the fishes

As accounted by Izaak Walton's Piscator in *The Compleat Angler*, '*The Salmon is accounted the King of freshwater fish*'. We might have considered this under the heading of 'Royal fishes' above, though the meaning here is more metaphorical than literal, pertaining to the sporting virtues and migratory habits of the Atlantic salmon, including the heroic efforts of mature fish to leap over obstructions in fast-flowing water. This charismatic species undoubtedly enjoys a high public profile, perhaps even the highest of all of Britain's freshwater fishes.

However, the Atlantic salmon also qualifies under the scientific definition of 'iconic species'. Like the Mahseer fishes of South Asia (in the genus *Tor*), sturgeon species of central Asia, giant stingrays of the Amazon and other large freshwater fish species, the Atlantic salmon is a top predator with a migratory lifestyle dependent on a range of linked habitats, all of which need to be adequately protected and connected for it to complete its life cycle. This scientific and figurative iconic status means that the Atlantic salmon is a suitable symbol of river protection and restoration. The viability of Atlantic salmon populations and of other characteristic fishes has played a significant role in the founding of a range of civil society Rivers Trusts across the UK mobilising public activism around conservation projects. Atlantic salmon are central to the charitable objectives of one of these Rivers Trusts: the Wye and Usk Foundation. Return of Atlantic salmon to the Thames system, where they were once abundant but became expunged through industrialisation, was the vision upon which the Thames Salmon Trust was founded in 1986, this voluntary organisation morphing subsequently into today's Thames Rivers Trust, dedicated to improving the river and its tributaries to benefit people and nature.

The iconic Atlantic salmon is also revered in continental Europe. The Gave d'Oloron is the longest salmon river in France, rising from a source high in the Pyrenees. The town

of Navarrenx on the Gave D'Oloron has been the epicentre of the World Salmon Fishing Championships for over forty years. On the Atlantic seaboard of northern Spain, some twenty-six rivers draining the Cantabrian range and the Galician Coast are run by Atlantic salmon, the species constituting a prominent feature in local festivals and folklore with salmon festivals timed with the run of these charismatic fish.

### 9.3.2 Christmas fishes

The common carp may not be native to Britain, though it is now widely established through multiple introductions over several centuries. However, in Germany, the common carp is common Christmas fare, as it has become in Hungary and a number of other northern European countries. In the 1817 book *Elements of the Natural History of the Animal Kingdom*, Charles Stewart states that '*In Germany it is taken from the pond, kept wrapped up in moss frequently wetted, fed with bread soaked in milk, and thus fattened for the table*'. Halászlé, a form of fish soup, is also traditional at Christmas in Hungary, with similar fish soups also featuring amongst festive fare in Poland and Serbia.

In Finland, whitefishes (Britain hosts two species) and zander are common constituents of Joulupöytä, a traditional Christmas smorgasbord. In Norway and Sweden, whitefish preserved with the white sauce lye that has been washed and boiled is known as *lutefisk* and has a traditional Yuletide place.

### 9.4 FISHY FACULTIES

We looked at the senses and faculties of fishes in the chapter 'Knowing your way round a fish'. However, some of these senses, sensory organs and brain functions warrant a little closer attention in this section on the curious world of Britain's freshwater fishes.

### 9.4.1 Memory like a goldfish

The saying 'Memory like a Goldfish' encapsulates an urban myth that goldfish have such short memories that they forget one end of a fish tank when swimming to the other. Science, however, exposes quite a different story. The myth of short-term goldfish memory was brought to wider attention in an experiment by an Australian schoolboy. Reported in many newspapers in February 2008, South Australian school student Rory Stokes from Adelaide, then fifteen years old, conducted an experiment conclusively demonstrating that goldfish, in fact, have very good memories.

Rory is reported to have undertaken his experiment to open people's minds to the cruelty of keeping these fish in confined bowls and small tanks. It entailed him teaching a small group of goldfish over a three-week period to swim to a beacon he had placed in the water. The conditioning entailed the schoolboy lighting the beacon at feeding time each day and, after a 30-second delay, sprinkling fish food around it. During the conditioning period, the goldfish took less and less time to swim towards the beacon as it was illuminated. In the first few feeds, they took more than a minute. After three weeks, this time shrank dramatically to less than 5 seconds.

**'Memory like a goldfish' may be an urban myth. (Image credit: tanuha2001/Shutterstock.com.)**

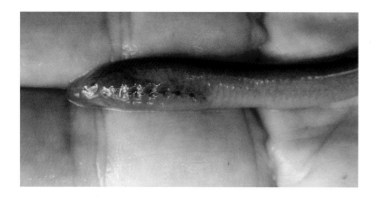

The six eye-like round gill pores behind the true eye on each side of the head of a lamprey and its ammocoete larva give these fishes the nickname 'seven eyes'. (Image © Dr Mark Everard.)

The beacon was then removed from the feeding process. However, after a period of six days, the goldfish responded in a mean time of 4.4 seconds after the beacon was again placed in the water and illuminated. This simple experiment provided that the goldfish retained the association between beacon and food. *'Memory like a Goldfish'* then may be something to which to aspire!

### 9.4.2 Fish eyes

Other fishes have eyes adapted to hunting in near darkness, including at night or in murky water. The zander is one such British (albeit introduced) freshwater fish. Looking at a zander, the fish appears to have 'dead eyes'. This is accounted for by the presence of a reflective layer behind the retina to maximise vision in near-dark conditions. This makes the zander a proficient hunter in low-light conditions.

Whilst on the topic of curious fishy facts concerning eyes, an ancient name for the ammocoete larvae of lampreys as well as the adult brook lamprey is 'nine eyes' or 'seven eyes'. For example, in the 1815 book *The Angler's Guide: Being a Complete Practical Treatise on Angling*, Thomas Frederick Salter writes, *'The lamprey eel is of the shape of the Lamprey or Seven Eyes… The lamprey eel is frequently caught in the river Severn, near Gloucester'*. In reality, all lampreys have a single pair of eyes, like most other fishes. The name 'seven eyes' refers to the perception created by the seven round gill pores on each side of the head behind the small eyes, superficially resembling additional eyes. (If we are entirely accurate about this, the seven gills and one eye on either side of the head should really give these small fishes the nickname 'sixteen eyes'!)

### 9.5 THE FRANKLY BIZARRE WORLD OF THE EUROPEAN EEL

All of Britain's freshwater fishes have unique traits, quirks and interactions with us terrestrial bipeds. However, were I pressed to call upon one species with the greatest natural and cultural curiosity, I think I would have to point to the European eel. In fact, so curious is the European Eel that it gets a chapter all to itself.

# THE CURIOUS
# AND CURIOUSER
# WORLD OF THE
# EUROPEAN EEL

This chapter is entirely devoted to celebration of this most curious of all British freshwater fishes: the European eel. It includes what we know about the eel, the tales we tell to fill in gaps in our knowledge, those stories we have repeated often enough to make us believe they are facts and a range of social curios about our relationship with this enigmatic fish.

## 10.1 THE MYSTERIOUS ORIGINS OF BABY EELS

Before the phenomenon of migration began to be understood, the appearance, disappearance and breeding of many organisms was a thing of mystery. People put their observations together to devise ingenious theories. Some were right or hinted at the truth. Others, with the benefit of hindsight, were fanciful.

One such widely held belief concerned the life cycle of the swallow (*Hirundo rustica*), that most graceful of avian summer visitors. As we all observe when spending time by fresh waters, swallows gather to swoop low over ponds, streams and damp meadows to hawk for insects in the autumn to feed up before then vanishing. The following spring, they appear suddenly and often in large numbers as a harbinger of spring, swooping and feeding voraciously on emerging insects over water. In the absence of contemporary knowledge, it is not such a misplaced intuitive leap to conjecture that swallows lie dormant throughout the winter in the very places they massed in autumn and reappear in spring, that is in the beds of large rivers and ponds. This view prevailed right up to the eighteenth century. It was also, as another of many examples, considered by many that geese (barnacle geese in particular, hence their common name), arriving suddenly in flocks, were born of the sun's heat on the long-necked and admittedly goose-like 'goose barnacles' washing ashore on drifting wood.

**The enigmatic European eel.**

**A group of tiny elvers climb the face of a weir to access upstream reaches of the river.**

Many such conjectures related to the sudden appearance of wriggling masses of tiny eels in the margins of lower reaches of rivers in the spring. Aristotle considered that this was because eels were born *'of nothing'*. This sudden appearance of small, hair-like elvers, often found in depressions made by horse hooves in river margins after rain, led Pliny the Elder (AD 23–79), the Roman author, naturalist and natural philosopher, to speculate that elvers grew from horsehairs dropped into the water or else that *'They rub themselves against rocks, and their scraping comes to life. Nor have they any other mode of propagation'*. Another prevalent view was expressed by Aristotle who, observing that there was no differences between the sexes of eels, considered that young eels were produced not by ova, but *'They are produced from what are called the entrails of the earth, which exist spontaneously in mud and wet earth'*. Aristotle proceeded to describe that there were places in the sea where putrefaction of seaweed and other such matter abounds.

In *The Compleat Angler*, Izaak Walton's Piscator noted that,

> But most men differ about their breeding: some say they breed by generation, as other fish do; and others, that they breed, as some worms do, of mud; as rats and mice, and many other living creatures, are bred in Egypt, by the sun's heat when it shines upon the overflowing of the river Nilus; or out of the putrefaction of the earth, and divers other ways. Those that deny them to breed by generation, as other fish do, ask, If any man ever saw an Eel to have a spawn or melt?

To this, Piscator adds,

*And others say, that Eels, growing old, breed other Eels out of the corruption of their own age;*
*which, Sir Francis Bacon says, exceeds not ten years. And others say, that as pearls are made*
*of glutinous dewdrops, which are condensed by the sun's heat in those countries, so Eels are*
*bred of a particular dew, falling in the months of May or June on the banks of some particular*
*ponds or rivers, apted by nature for that end; which in a few days are, by the sun's heat, turned*
*into Eels: and some of the Ancients have called the Eels that are thus bred, the offspring of Jove.*
*I have seen, in the beginning of July, in a river not far from Canterbury, some parts of it covered*
*over with young Eels, about the thickness of a straw; and these Eels did lie on the top of that*
*water, as thick as motes are said to be in the sun: and I have heard the like of other rivers, as*
*namely, in Severn, where they are called Yelvers; and in a pond, or mere near unto Staffordshire.*

The earliest known scientific work on eel biology, *Libri de piscibus marinis in quibus verae piscium*
*effigies expressae sunt*, was written in Latin and published in 1554 by Rondeletius. The work was
subsequently translated into the French of that era with the title *L'histoire entire des poissons* (*The*
*entire story of fish* in English). Rondeletius, or Guillaume Rondelet (1507–1566), was a professor
of medicine at the University of Montpellier in southern France. Rondeletius became renowned
as an anatomist and a naturalist, including this major work on marine animals that was to become
became a standard reference work for about a century. A large part of the work is devoted to eel
biology, noting of the absence of visible gonads in adult eels, that they were likely present but were
hidden by fat. Rondeletius, as it happens wrongly, described eels of different sexes despite the
lack of gonads to corroborate his account, his words translated as, '*In Languedoc, southern France,*
*there are two types: the male, which has a shorter but larger head, is referred to as* margaignon; *the*
*female, which has a smaller, more pointed head, is referred to as anguille fine [literally 'thin eel']*'. Of
their breeding, Rondeletius then continues, '*Each eel gives birth in freshwater, and it is the only one*
*of all similar fish that enters marine lakes, including the sea, but otherwise lives in rivers, lakes and*
*ponds*'. That eels have an affinity for both saline and fresh water is true, even if Rondeletius was
incorrect about them breeding in fresh waters, as well as about morphological differences between
the sexes when in their freshwater stage. He also incorrectly describes eels as being scaleless (as
did Izaak Walton), when in fact eels have fine, inconspicuous scales embedded in their skin. Many
other authors noted a difference in the structure of European eels at different life stages (a point to
which we will return later), a number of these people suggesting that these were different species.

These various myths were abandoned as the phenomenon of migration was discovered,
and various strands of evidence around the eel's complex life cycle coalesced. However, the
probable reality of the eel's life cycle is not only a complex and seemingly improbable tale, but
also one in which received wisdom still includes a number of sweeping assumptions papering
over major evidence gaps.

## 10.2 WHAT WE KNOW AND THE STORIES WE TELL

It is easy to denigrate our previous generations for beliefs held in ignorance of today's scientific understanding and received wisdoms. However, this would be to blind ourselves to the tales we still tell to fill in gaps for which we have no evidence. In the simple telling of the life cycle of the European eel in the chapter *Sex lives of the British freshwater fishes*, I implied that all was not quite as simple as received wisdom would have us believe. In reality, the life cycle of the European eel is even more remarkable than the various abandoned myths.

Marine scientists had long ago discovered all manner of small organisms floating as plankton in oceanic surface waters. One such group of broadly similar small (less than 5 mm long), laterally flattened and transparent leaf-shaped fishes was assigned to the genus *Leptocephalus*, literally meaning 'small head'. Today, we know of sixteen different families of leptocephalus organisms, including over seventy different species of fish. It was also known that no eel had ever been found with testes (male reproductive organs).

It was the Danish biologist Johannes Schmidt (1877–1933) who, in 1920, connected the strands of knowledge. Schmidt discovered that, although they look nothing like adult eels, leptocephali eventually metamorphose into tiny slim, transparent eels known as elvers or 'glass eels' by the time they reach European coastal waters. European eel leptocephali have been found drifting passively in the Gulf Stream from the edge of the Sargasso Sea, some 5,600 km (3,480 miles) away from our shores in the western Atlantic. As the leptocephali drift, they feed and grow, and are also fed upon by many other organisms. On completing their long journey

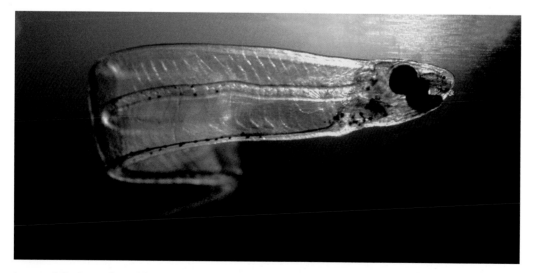

Leptocephalus larva of an eel (this one of a 7.9 mm long marine conger eel leptocephalus rather than of an otherwise similar freshwater European eel). Leptocephali were once considered a separate marine planktonic group of species. (Image source: Kils at the English language Wikipedia, CC BY-SA 3.0, used under https:// creativecommons.org/licenses/by-sa/3.0/deed.en.)

of around two years as they enter coastal water, the leptocephali then metamorphose into a small, transparent recognisably eel-like body form known as a 'glass eel'.

On arrival at the shore, the glass eel stage of the European eel then funnels into estuaries, gradually taking on darker body pigments to then become known as elvers. Many of these elvers then penetrate up into freshwater systems from North Cape in Northern Norway southwards along the coast of Europe and the Mediterranean, including the rivers of the North African coast. Not all of these eels enter fresh waters, many remaining in estuaries and coastal seas.

Once in fresh waters, European eels can live for anything from six to thirty years, attaining a length of between 40 and 100 cm long. In this freshwater adult phase, the mouth is broad, the eye small and the body grey, often with a slightly yellow tinge, hence the common term 'yellow eel'. However, when they mature, the eels then metamorphose once again into 'silver eels' with a slimmer profile, wider head and smaller mouth, larger eyes and a more silvery colour. Silver eels then migrate back to sea, down rivers or from still waters into rivers. This can often entail snaking through moist vegetation from isolated pools. These downstream migrations occur particularly in darkness on warm autumnal nights, often under a new moon and favoured by a rising river.

This propensity of large eels to migrate downstream during autumnal storms was noted by the Roman natural philosopher Pliny the Elder, who observed the abundance of eels caught during stormy weather in October, as it had been by the ancient Greek philosopher and scientist Aristotle (BCE 384–322), who believed this was due to eels having only small gills and so therefore likely for them to suffocate when the water was muddy. English zoologist William Yarrell (1784–1856), in his 1836 two-volume book *The History of British Fishes*, comments that basket fishermen in the tidal reaches of rivers placed their eel-pots facing upstream in the autumn and downstream in the spring. Strands of evidence were coalescing that breeding may occur at sea, with the tiny elvers returning upstream in the spring.

**The silver eel form of the European eel. (Image © Nicolas Primola/Shutterstock.com.)**

The instinct to return to sea is very strong, these rugged fish eventually entering estuaries and taking to sea, where other eels not entering fresh waters have lived out their lives. It is speculated that sexless leptocephali entering fresh waters later develop into females, and those in salty water become males. But, like many pieces of received wisdom concerning the European eel, this is really pure speculation as no one has ever observed a European eel with male gonads.

It is then assumed that the maturing silver eels make the long crossing back across the Atlantic, ceasing to feed and changing their metabolism to devote resources to producing eggs and sperm for reproduction, with the female eels heavy with eggs by the time they reach their spawning grounds. However, although the drift of leptocephali from the border of the Sargasso Sea is well evidenced, there is not a shred of direct evidence that mature eels return there to spawn. Quite where spawning occurs is in fact another of nature's great mysteries. To unravel the secret of silver eel migration, scientists have inserted biodegradable satellite tags into mature eels released from the coast of Galway in Ireland. These miniaturised, pop-up satellite tags collect data concerning the eel's daily activity, including the speed, depth and direction of their movement. The biodegradable link eventually degrades and breaks down after around six months, the tag then bobbing to the sea's surface to transmit data to satellites from which it is then received by scientists. However, data thus far recovered reveal only what these eels have been doing in the first 1,300 km of their epic journey. The eels released from Galway initially took a southerly course, swimming near the surface by night but dropping down to depths of around 1 km by day. This behaviour suggests both avoidance of light and presumably sight-feeding predators, but also that they may have been heading into the deeper 'conveyor belt' ocean current eventually leading towards the Sargasso Sea. However, the rest of the 'knowledge' we have about where eels go and what they do leading to the generation of leptocephali is pure speculation. It is simply a story we tell ourselves in the absence of hard evidence but that is received as truth through constant re-telling, including in numerous books, some of them illustrated with drawings of eels spawning in floating weed mats in the Sargasso Sea. The hard reality is that no eel has ever been found in the Sargasso Sea, nor has any European eel yet been found with testes.

The Sargasso Sea itself is another oddity. Firstly, it is the world's only 'sea' without a shore. The Sargasso Sea is also known as the North Atlantic Subtropical Gyre, spanning an area of 1,100 km (700 miles) by 3,200 km (2,000 miles). It is bounded not by land but by strong oceanic currents including the Gulf Stream to the west, the North Atlantic Current to the north, the Canary Current to the east and the North Atlantic Equatorial Current to the south. Lacking a coast, the Sargasso Sea is highly oligotrophic (lacking in nutrients), and so the water is very clear and its plankton sparse. This makes it far from an ideal place for the young of fish of any species to thrive. Near the Bermuda Triangle, the Sargasso Sea is also extremely calm. With little wind, ships could be becalmed for days, believed to account for the demise of many sailors in a 'sea of lost ships' in historical accounts. However, the great depth of the Sargasso Sea and the long penetration of light in its nutrient-poor waters does mean that there is plankton at considerable depths where upwelling currents introduce nutrient chemicals, possibly presenting suitable conditions for spawning and larval growth.

Given the complexity and seemingly implausibility of the life history of the European eel, and that no one had ever seen breeding of eels nor any with testes, it is then hardly surprising that our forebears held all sorts of strange ideas about the origins of baby eels. But hard facts are as yet in short supply to support or disprove what many people accept to be the truth.

### 10.3 WHEN IS AN EEL NOT AN EEL, OR AT LEAST NOT THE EEL WE THOUGHT IT WAS?

We have already seen how leptocephali were once considered entirely different organisms than adult eels, grouped into a now 'wastebasket taxon': the genus *Leptocephalus*. In Izaak Walton's day, various other divisions in eel 'species' were described in *The Compleat Angler*,

> *And lastly, let me tell you, that some curious searchers into the natures of fish observe, that there be several sorts or kinds of Eels; as the silver Eel, the green or greenish Eel, with which the river of Thames abounds, and those are called Grigs; and a blackish Eel, whose head is more flat and bigger than ordinary Eels; and also an Eel whose fins are reddish, and but seldom taken in this nation, and yet taken sometimes.*

Based on current knowledge, we now regard all as different morphs throughout the life cycle and as localised adaptations to prevailing conditions of the single European eel species.

The Reverend Houghton identified two distinctive species of eel in his 1879 book *British Fresh-water Fishes*. Houghton distinguished the broad-nosed eel (*Anguilla latirostris*) from the sharp-nosed eel (*Anguilla vulgaris*) largely on head shape, discounting alternative prior divisions into the three species: sharp-nosed eels (*Anguilla acutirostris*), middle-nosed eels (*Anguilla medirostris*) and blunt-nosed eels (*Anguilla latirostris*). Henry Cholmondeley-Pennell, in his 1863 book *The Angler Naturalist*, considered that the broad-nosed eel does not migrate, compared with other eel species that did. Houghton concluded that his broad-nosed eel was the same species as the frog-mouthed eel reported by fishermen on the River Severn. Houghton also noted that the fishermen of Toome, a town on a tributary of Lough Neagh, paid their rent in eels, but that broad-nosed eels were thrown back in the water.

Once the general principles of the life cycle of the European eel had been established, a belief arose that different types of eel from all over the world migrated to the Sargasso Sea to breed in one giant melting pot. The theory built on discovery of the movement of tectonic plates,

**The so-called sharp-nosed eel and broad-nosed eel, described by the Reverend Houghton and painted by Alexander Francis Lydon.**

eels breeding in a favoured pool that widened as continents divided. We now know that the Indian mottled eel spawn in the Indian Ocean, though precisely where remains a mystery as, compared with other species, very few Indian mottled eel leptocephali have been captured. This may be due to the species having single or multiple spawning locations, which may overlap with those of other eel species. Australian eels, including the speckled longfin eel (*Anguilla reinhardtii*, also known as the Australian long-finned eel) and the short-finned eel (*Anguilla australis*), spawn in the Coral Sea near New Guinea. Their tiny sexless leptocephalus larvae develop as they drift southwards on ocean currents for between one and three years before entering lower Australian rivers. At least for Australian eel species, it has been discovered that their subsequent adult habitat determines their sex, larva that travel upstream into fresh waters becoming females whilst those that stay in estuaries develop into males.

What remains more of a mystery is how the European eel differentiates itself from the American eel (*Anguilla rostrata*). These two eel species are the only ones of the sixteen known eel species that inhabit the Atlantic, all other eel species occurring in the Indo-Pacific, perhaps reflecting the evolutionary origins of the family. According to received wisdom, both the European eel and the American eel spawn in the Sargasso Sea, various scientific papers assuming that they have separate but partially overlapping spawning sites. The leptocephali of European eels have been scientifically demonstrated to drift on the Gulf Stream to European shores, whilst those of the

American Eel drift a substantially shorter distance on the North Atlantic Drift to American shores. Many experts were of the opinion that these juvenile eels developed subsequently into different 'species', according to differing environmental conditions where the currents took them. However, subsequent DNA analysis, as well as the fact that European Eels have more vertebrae (bones in the backbone) than American eels, has determined that these species are quite distinct. However, hybrids between the two Atlantic species occur in Iceland, accounting for around 15% of the population there, with European eels also found there but not pure strains of the American eel. Hybrids are not known elsewhere, only in Iceland where glaciation during the Pleistocene meant that freshwater habitats were too inhospitable to support eels. (In fact, a very few hybrids have been found in eel populations in rivers running to the Mediterranean, but these are scarce enough to suggest that they are the product of transfer of glass eels imported from Iceland.) How, then, could it be that hybrids between these species only occur more or less where their adult ranges overlap if the assumption holds true that these fishes spawn thousands of miles away in the Sargasso Sea? We have to recognise that received wisdom about the missing evidence links in the life cycle are in reality fabrications to explain what happens between maturing eels migrating and leptocephali larvae appearing in oceanic currents. Various theories have been advanced about the differential survival of pure strain and hybrid leptocephali on oceanic currents to explain this oddity. In my mind, a more localised spawning site may provide a more convincing explanation. As ever with the European eel, increasing knowledge only seems to intensify the mystery.

Truths of the European eel's life cycle that are revealed by science are bizarre enough, perhaps more so than the long-abandoned myths. But, even in the absence of supporting facts, we remain loyal to the stories we tell ourselves about major missing elements of evidence.

### 10.4 EEL FISHERIES

Although the life cycle of the European eel was known only in fragments with a fascinating range of fabrications erected to explain the gaps, as indeed we still do today, the springtime upstream migration of glass eels and elvers and the autumnal downstream migration of adult eels were well known. These predictable migrations were the key to many types of eel fisheries.

Some forms of eel fishing target non-migrating stages in the life of the adult eel. An early and very simple device for eel fishing was the eel spear. Aristotle mentions a three-pronged spear used to catch eels and also flatfish, its description similar to the three-pronged eel spear still in common usage in the Reverend Houghton's day. A very few eel trappers today still operate on the Fens in Cambridgeshire, using traditional designs of wicker pots with inverted funnels in their necks or else similar designs made out of modern materials. However, unbroken histories of some eel-catching families date back over 500 years, and archaeological evidence from the nearby ancient Fenland town of Whittlesey suggests that eel catching and its way of life could date back to the Bronze Age 3,000 years ago, as almost

**An eel spear, an ancient device used to hunt for eels. (Image © Allan Frake.)**

identical sets of traps and remains of boats have been widely discovered in Britain. Angling too is a modern form of eel fishing, though nowadays almost exclusively recreational, with fish returned alive to the water.

Glass eel fisheries are based in tidal rivers where these tiny eels arrive on European shores before colouring up to turn into elvers. The principal areas of glass eel fishing are on the west coasts of Spain, France and the UK, with some limited fishing in the Mediterranean Sea coasts of Spain, France and Italy. The start of the glass eel fishing season is highly dependent on the prevailing weather. The UK glass eel fishing seasons runs from February to May, the bulk of the catch generally occurring during April, though some glass eels are caught as early as January and as late as June. However, the timing of the glass eel run varies considerably from year to year, with substantially reduced catches during chilly weather and floods. The majority of British glass eel fishing occurs on the River Severn system. The Sustainable Eel Group estimates that approximately 75% of the total European glass eel catch is taken in France. Methods vary too. Most glass eels are caught by small trawlers using 'wing nets' and 'trawls' in France. The only permitted instrument for glass eel fishing in the UK is a traditional form of hand net with a long handle and a bottom width of six feet (2 m), a device used for hundreds of years to scoop up glass eels migrating upstream in the river's margin. Most of the captured glass eels are transported to international markets, generally in thermally insulated containers suitable for loading into aeroplanes.

Eel traps have also been commonly used in rivers throughout history. Eels were taken in wicker baskets with narrow necks in Roman times, as also in baited earthenware vessels covered in colander-shaped lids. On major river systems such as the Hampshire Avon, many weirs and

Glass eels.

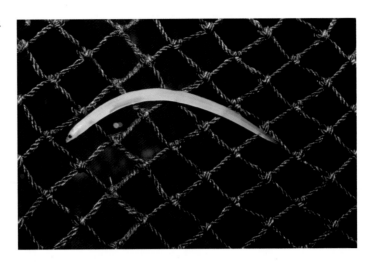

An eel trap built into a weir on the Hampshire Avon at Winkton, activated by opening sluices to divert river flow through the grid, now no longer routinely used because of conservation concerns but formerly harvesting European eels as they migrated downstream on autumnal spates. (Image © Dr Mark Everard.)

former mills had eel traps built into them. Particularly on warm, dark and stormy nights with a new moon in October, water would be diverted through grated eel traps, the migrating eels gathered up for collection and sale. These traps are largely not in use today, mainly because of increasing conservation concerns about the species but also their relative scarcity making trap operation more a tradition than a business. However, some of these traditional eel trapping operations persist where permitted by historic licences.

For all its fondness for a dish of them, London was reported in the Reverend Houghton's 1879 book as chiefly supplying Holland with eels. Today, the dominant export market is to Japan, where eels, known as unagi, are a great and expensive delicacy.

## 10.5 PET EELS

When I was a boy, I had a pet eel called Eric. Eric was caught from a local stream and spent half its first night in captivity in an aquarium. The second half of the night was spent crawling around the shed floor and finding refuge in a musty corner, judging by the slime trail I found in the morning. Eric was duly returned to the water somewhat dried up but otherwise unscathed, and I tied some fine netting over the top of the tank. He stayed with me for quite some time after that.

Other eels have had more august keepers. Earlier in this chapter, I related what Francis Bacon had recorded about Crassus the orator and also Hortensius lavishing affection on their lamprey pets. However, this raises questions about how exactly a parasitic species was fed during its reported long life in captivity. It would appear that this is, however, a result of mistranslation of the Greek and Latin term 'murena', applying equally to lampreys and eels. European eels are by far the more likely residents of these early fish ponds and are also an intelligent species capable of responding in this way to people. (A friend of mine tells a tale of how a lady neighbour used to bang on a metal post in the river margin to call to an eel, which it had learned was a signal that she had some titbits of food to offer it.)

Other surviving ancient Roman texts record that Gaius Hirrius was the first person to have ponds solely for raising eels, and to have supplied Caesar with six thousand of them for his triumphal banquets. The orator Quintus Hortensius, Consul in the Roman Empire in 69 BCE, was said to have lavished great care on his fishpond and to have wept on the death of his favourite *murena* (now presumed an eel unlike the reports by Francis Bacon that it was a lamprey). Augustus, considered to be the first Roman emperor, having controlled the Roman Empire from 27 BCE until his death in AD 14, had a niece called Antonia. Antonia was another doting eel owner, adorning the fish with earrings (hinting that this was a European eel as lampreys and marine moray eels lack pectoral fins or gill covers on which to hang these ornaments). Crassus the orator, too, was said to have adorned a pet eel with earrings and small necklaces '*just like some lovely maiden*', training it to respond to its name and swimming up to eat what was offered. Lucius Domitius Ahenobarbus (Consul in 54 BCE) mocked Crassus for weeping upon the death of his much loved pet.

## 10.6 EELS AND PSYCHOANALYSIS

As we have seen, no one has ever found testes in a European eel. This is clearly something of an enduring mystery and has been the subject of substantial research interest in former times. One such early investigator was Sigmund Freud. Freud is, of course, far better known as the Austrian founder of the psychoanalytic school of psychiatry. However, the outset of his long career entailed research into the mysteries of eel reproduction.

In 1876, at the age of just nineteen, the young Freud travelled to the University of Vienna's zoological station in Trieste, Italy. This research station was situated on the coast of the Adriatic Sea. It is here that Freud set about dissecting eels, seeking to discover their testes. However, very many dissections later, the young Freud was no more successful than the many other anatomists who had preceded him. Freud was to write to his colleagues that '…*all the eels which I cut open are of the fairer sex*'. Freud's first scientific paper was titled 'Observations on the Form and the Finer Structure of the Lobular Organs of the Eel, Organs Considered to be Testes'. The research in reality only served to deepen the mystery of eel reproduction. Freud famously then abandoned biology in favour of working on the human mind, proceeding to become the 'Father of Psychology'.

**Sigmund Freud, to become the 'Father of Psychology', started his scientific career dissecting European eels in the vain quest of finding one with testes.**

## 10.7 EEL GASTRONOMY

In considering the roles that British freshwater fishes play as food, we have already celebrated jellied eels and 'eel pie, mash and liquor' as traditional dishes, particularly in London and southeast England, and also the popularity of smoked and baked eels in continental Europe. In *The Compleat Angler*, Izaak Walton also notes that '*It is agreed by most men, that the Eel is a most dainty fish: the Romans have esteemed her the Helena of their feasts; and some the queen of palate-pleasure*'.

Eels, and glass eels in particular, are a very highly prized delicacy across Asia. Indeed, they are so prized that the glass eel stage of the European eel commands hundreds of pounds sterling per kilogramme, prices rising with declining eel populations. Glass eels are particularly valuable, both consumed directly but also as stock to be grown on into adult eels, increasing their weight and market value substantially.

Unagi is the Japanese word for the freshwater eel, relating specifically to the Japanese eel (*Anguilla japonica*) but also used more widely for other eels, including those imported as food. Unagi is a common ingredient used in Japanese cooking, served sliced on a bed of rice, as a kind of sweet biscuit called 'unagi pie' made with powdered unagi, as *unakyu* (sushi containing eel and cucumber) and as a prized addition to other dishes. So popular is the eel in Japan that specialised unagi restaurants are common, often identified by signs showing the word *unagi* with the Japanese hiragana script, or more commonly the hiragana character that translates simply to 'u'. Lake Hamana in Hamamatsu city, Shizuoka prefecture, is thought to produce Japan's highest quality unagi, and so, consequently, the lake is surrounded by many small restaurants that specialise in unagi dishes. Because of health hazards associated with

Jellied eels are a traditional and popular dish in some localities, particularly in London. (Image credit: Guy Erwood/Shutterstock. com.)

eating of raw freshwater fishes, eels are always served cooked, and often served with tare sauce in Japanese food. So great is their hunger for eels that Japan consumes more that 70% of the global eel catch.

## 10.8 THE ENDANGERED EEL

Since 2014, the European Eel has been classified on the IUCN 'Red List' as Critically Endangered, largely due to the steep rate of decline of the species across its range. European eels used to thrive in British rivers, a staple and traditional part of the diet, particularly in London and southeast England, and supporting productive fisheries and export markets elsewhere across the nation. However, there has been an estimated population decline of 90%–95% since the 1980s, triggering a range of nature conservation responses. Conservation concern and action must necessarily be international, reflecting the complex life cycle of the species and its wide European range, even extending into northern Africa.

The European Commission initiated an Eel Recovery Plan (Council Regulation No.1100/2007) to try to return the European eel stock to more sustainable levels, including a range of measures concerning stocking with eels. Under this 2007 Regulation, at least 35% of the commercial catch of eels of less than 12 cm long had to be made available for restocking, rising to 60% by 2013. The Environment Agency led the development of a national Eel Strategy, guidance on eel and elver passes over obstructions and regional strategies including, for example, a Thames Eel Management Plan. These regional management plans outline key targets and measures to improve the population and conservation of the European eel. Given the scale of the challenges faced by eels, working partnerships amongst concerned institutions is essential.

British responses include Citizen Scientist monitoring of eels coordinated through bodies such as the Zoological Society of London, as well as the Rivers Trusts which is leading the primary school 'Eels in the Classroom' initiative considered later in this chapter. The Sustainable Eel Group has been a champion of the European eel, bringing bodies together to develop workable strategies. More formal, scientific surveys of eel populations are also being undertaken. Many bodies are collaborating also to improve migratory pathways for eels, including installation of eel passes on weirs or else weir removal where these obstructions currently fragment eel populations.

Daunting challenges face the European eel population, with threats right across the many linked habitats essential for them to complete their life cycles. However, conservation responses have at least taken root and are being implemented by a range of connected organisations, both statutory and voluntary, and a network of concerned individuals.

## 10.9 THE SHADY WORLD OF EEL SMUGGLING

With the phenomenal prices commanded by eels, glass eels in particular, on global markets, the strictures on eel fishing and the increasing scarcity of fish to catch, it is hardly surprising that there is a substantial black market in the illicit capture and trade in eels.

In June 2017, UK Border Force officials apprehended a 64-year-old man at Heathrow Airport and, beneath a crate containing chilled fish legitimately bound for Hong Kong, found a package of 600,000 glass eels with an estimated street value of £1.2 million. This is a very high-profile case illustrative of a substantial and booming illicit trade particularly across Europe. Were eels as abundant now as they have been historically, when major rivers in the west of England glittered on rising spring tides as elvers migrated upstream *en masse* and adult eels comprised around half of the biomass of all freshwater fishes, there would not be such a hue and cry. However, with the massive collapse in Britain's population of European eels over the past three or four decades, the very survival of the species is at stake. Glass eel fisheries are now very tightly controlled across Europe, and of those little fishes that are netted in the February to May fishing season, approximately half is taken for human consumption and the other half used for restocking across the continent. Legally, all exports have also been banned to Asia, where the eel is highly prized for its purported strength-giving properties.

In the shadow of this decline in eels stocks and their legitimate trade, a vast and growing black market is thriving involving highly organised elver smuggling rings using techniques such as specially adapted suitcases to smuggle the fish out to Asian markets. Profits in eel smuggling are of the same order of magnitude as those for illegal drugs. Once finding their way by nefarious routes into Asia, the elvers are grown on in specially created lagoons, their market value increasing by as much as ten-fold within a year. The sheer scale of illegal eel smuggling is today threatening the survival of the species.

## 10.10 EELING AND HEALING

Medicinal qualities have been attributed to the European eel since antiquity. Izaak Walton writes of it in *The Compleat Angler*: '*And I will beg a little more of your attention, to tell you, that Aldrovandus, and divers physicians, commend the Eel very much for medicine, though not for meat*'. Houghton records that eel skin was a good remedy for cramp or rheumatism. Other eel-related remedies are reported in the chapter, 'What have freshwater fishes ever done for us?'

A recent beauty treatment fad has been that of exfoliation by immersing a person's body in a bath full of live elvers. Whilst the eels are reputed to be efficient in nibbling off loose skin flakes as a form of food, problems have perhaps unsurprisingly arisen for those wearing loose-fitting swimwear. There are reports of one elver in China in 2011 wriggling its way up the penis of a man called Zhang Nan, up through his urethra and entering his bladder. The poor man had to endure a three-hour operation to remove the six-inch long eel, obviously dead by the time doctors found it. Though the urethra is narrower than the eel's body, such is the agility and slipperiness of the elver that it was able to make a smooth entry. These are not the only reported instances of other freshwater fish species entering the urethra in fresh waters in other countries, but I will stop here, not merely as it is getting a little off the focal point of this section, but also to protect the squeamish.

## 10.11  EEL POISONS

The potentially dangerous properties of eels were well known in Izaak Walton's day, his Piscator noting,

*But now let me tell you, that though the Eel, thus drest, be not only excellent good, but more harmless than any other way, yet it is certain that physicians account the Eel dangerous meat; I will advise you therefore, as Solomon says of honey, 'Hast thou found it, eat no more than is sufficient, lest thou surfeit, for it is not good to eat much honey.' And let me add this, that the uncharitable Italian bids us 'give Eels and no wine to our enemies'.*

Eel blood is poisonous to humans and other mammals, although both cooking and digestion destroys the protein responsible for its toxicity.

## 10.12  EEL AND EDUCATION

The very existence of European eels, their fascinating life cycles and the diverse and curious facts about them constitute a great educational resource. However, the 'Eels in the Classroom' programme led by the UK Rivers Trusts takes eels right into primary schools. A small proportion of elvers caught commercially are donated for use in this programme, with around 100 tiny eels stocked into fish tanks in each participating school. The children in the class then rear the juvenile European eels for the term, feeding them every day and learning about all aspects of their lives. At the end of five weeks, each school then has a release event, the children participating in the release of eels into their local river.

This is not only fascinating and engaging for the children but it opens their minds, prompting many questions concerning the life cycle of the eels, where they come from, what they eat, how they communicate and many more. The programme also provides a live learning environment about current environmental issues behind the reclassification of the species as Critically Endangered as a reaction to the numbers of European eels reaching Europe. The Eels in the Classroom programme has inspired children to take action for rivers in the future, particularly those in their neighbourhood that are now the home of 'their' released eels, as well as the wider health of all wildlife that depends on the river and the marine environment.

'Eels in the Classroom' offers environmental education to school children and a sense of responsibility for the rivers into which 'their' eels are eventually released.

Another anachronistic use of eels in 'teaching' was reported by Pliny, from ancient Rome. Eel skins were said to be used to whip naughty boys. The practice clearly persisted into the fifteenth century, when François Rabelais, the French Renaissance writer, penned, *'Whereupon his master gave him such a sound lash with an Eel-skin, that his own skin would have been worth nothing to make bag-pipe bags of'*. The Reverend Houghton reports that the term 'anguilla' was in later times applied to a whip made of leather thongs used to flog boys. Fortunately, such 'lessons' are now illegal!

## 10.13 EELS AND LANGUAGE

The derivation of the ancient Greek term for 'eel', which I am not even going to try to replicate here let alone try to pronounce, is said by Rondeletius to mean 'from silt'. This is plausible, given that silt is one of the favoured habitats of the European eel. The Latin *Anguilla* (also the name of the genus in the Latin binomial system) is said to be after the resemblance of eels to snakes, or *Anguis*, from which the French word for eel *anguille* derives.

The sinuous eel has also inveigled its way into our common parlance through terms such as 'slippery as an eel', and the same sense when the noun 'eel' is used as a metaphor for something or someone that is slippery. There are also many quotes of varying antiquities and degrees of seriousness. From Leonardo da Vinci, we have, *'Marriage is like putting your hand into a bag of snakes in the hope of pulling out an eel'*. Bertrand Russell gave us the line, *'Love is a slippery eel that bites like hell'*. And who can forget the Monty Python sketch on helpful English phrases for foreigners: *'My hovercraft is full of eels'*.

The town of Ely in the Fens is said to be derived directly from the abundance of eels, as are Elmore in Gloucestershire and Ellesmere in Shropshire. So too Eel Pie Island, the island in the

**The 'Isle of Ely', said to be named after the abundance of eels around this elevation of land in the surrounding wetlands. (Image credit: Peter Moulton/Shutterstock.com.)**

River Thames at Twickenham, London, known as a major jazz and blues venue in the 1960s, that took its name from the eel pies served by the hotel on the island in the nineteenth century.

### 10.14 EELS AND RELIGION

Gastronomy and religion have run head-to-head into each other regarding the eel. The King James version of the Old Testament contains text outlining that it is acceptable to eat 'fin fish', but fish lacking fins are an abomination. Of course, eels have fins, albeit lacking ventral fins, but distinctions arise when considering fish with and without scales.

For Jewish people, eels are not considered kosher, under which code fish have to have both scales and fins. In biological terms, eels clearly have fins and they also have scales albeit that the scales are small and tightly embedded in the skin. (Both Rondeletius and Izaak Walton wrongly stated that eels lack scales.) The situation is clarified under Torah law, the Jewish law of Kashrut, which states that '*The scale must be easily removable without tearing the skin*'. As the scales of the eel are embedded within the skin, the eel falls under the category of 'skin fish' (like lampreys and many catfish) rather than 'scale fish', and so is not kosher.

Under the Islamic world view, the status of eels as halal (allowed) or haram (prohibited) food is less clear. Under the halal code, the fish must come out of the water alive. Contention then arises, along the lines of Torah law, as to whether the eel is a 'scale fish', thought to release poisons into the water between its scales, or a 'skin fish' considered unclean as it accumulates poisons. Islamic law is less dogmatic on this point with varying interpretations, though as the eel is primarily a carnivore, albeit with strong omnivorous tendencies, this means that the fish is theoretically haram.

## 10.15 EEL AND THE ARTS

Eels of various types have featured in art over centuries, from antiquity to contemporary. Perhaps it is the snake-like form and the associated symbology of these fishes that makes them such popular subjects?

In ancient Egypt, Anubis had an associated eel that was both an incarnation of the God but also a carrier for a stage in the final underworld journey of the soul. Other cultures, such as in Oceana, portray the eel as a fertility symbol. In many of these island cultures, as in New Zealand, the lack of snakes has resulted in the superficially snake-like eel filling in the iconic niche filled by serpent symbolism elsewhere. For example, in Tonga, an eel seduces the virgin Goddess Hina in a tale similar to the roles of the serpent in the Biblical tale of Eve and Adam's fall from grace.

In the eBook *Charms, Spells, and Curses*, Victor Banis mines some of the mystical folklore around the eel, offering,

> *The eel has many marvelous* [sic.] *virtues. Let him die out of the water and steep his body in strong vinegary and the blood of a vulture, and place the whole under a dunghill. The composition will raise from the dead whatever is brought to it and give life as before. Also, anyone who eats the still-warm heart of an eel will be seized with a spirit of prophecy and will be able to foretell future events.*

I am unsure if the author tested any of these methods! In fact, as it is a cold-blooded animal, it is questionable exactly how one could eat *the still-warm heart of an eel*'.

## 10.16 EEL CONSERVATION

We will turn to the wider picture of conservation of Britain's freshwater fishes in the following chapter. However, specific efforts targeted at the European eel are significant and worthy of attention here, responding to the precipitous decline in eel numbers over recent decades.

The Sustainable Eel Group (SEG) was set up in 2009 and now operates through a network of one thousand contacts across Europe. SEG seeks to understand why the species has experienced such a marked decline and to accelerate its recovery. SEG activities include leading efforts to bring eel smuggling under control across Europe and monitoring the extent to which gangs are trying to infiltrate Britain. SEG has been instrumental in attracting commitments of substantial funds, most of which is being used to remove obstacles along waterways preventing elvers travelling upstream.

Somerset's River Parrett has become a model of how SEG aims to operate across Europe. Here, science, conservation and the eel fishery industry have combined to work together to promote the survival of the species. To bring transparency to the market for glass eels, elvermen sell their catch on to licensed buyers, who set up for the night near the riverbank where glass eels are caught. This market is highly regulated, unlike markets across the wider range of the species. Concerted efforts are required to protect the young eels at this vulnerable stage in their life histories if the species is to survive.

**Netsman working the margins of the River Severn to capture glass eels entering the estuary by night.**

Evolution over 3.85 billion years has imbued the natural world with bewildering biological diversity, a high degree of embedded resilience and the capacity to adapt. This has been through the process of natural selection, in which of the genes of the most fit to the diversity of habitat and niches within the natural world are preferentially passed on to the following generation. The most fit include adaptation to availability of food, interactions with other co-evolved species and many other aspects of the wider ecosystems of which they are part. This process meshes organisms in with the 'checks and balances' of predators, diseases, and environmental variability and stresses, resulting in tightly interlinked ecosystems, which themselves are elegantly adapted to environmental conditions.

Individual fish species are integral parts of wider fish communities, fish communities are intimately interlinked with whole aquatic ecosystems, and aquatic ecosystems are wholly interdependent with the wider drainage basins and connected coastal seas and climatic systems of which they are part. As mentioned on a number of occasions throughout this book, the conceptual disconnection between fishes and other wildlife makes no sense in terms of the tight interactions within ecosystems and the parts fishes play in food webs, as prey, consumers, hosts to parasites and their many roles in wider ecological processes and the ways people depend upon them.

## 11.1 VALUING BIODIVERSITY

The term 'biodiversity', shorthand for 'biological diversity', is now in common usage. However, the multiple dimensions of diversity are perhaps less generally understood.

Number of species alone is just one part of biodiversity, though not necessarily its best indicator. Some habitats are naturally species-poor, including fish-free turbulent headwaters. The 'trout zone' of an upper river may contain just juveniles of trout and salmon as well as bullheads sheltering under those stones that are not too frequently disturbed by spate flows. Rivers, pools, lochs and loughs in formerly glaciated regions may have an equally sparse natural fauna. Isolated ponds may become well vegetated with diverse invertebrate and amphibian fauna, but may naturally lack fish. Introducing fish and other species into these environments can have, and has had, adverse effects on native species and ecological balance. More important than sheer numbers is the appropriateness of species present, be that a naturally impoverished fauna or else multiple species adapted to diverse niches. For example, the fish fauna of lowland rivers may include common bream, roach, silver bream, pike, perch, chub, three-spined sticklebacks, bullheads, minnows, spined loach, ruffe and many more in various depths between the river channel and its margins and in connected wetlands.

Another important aspect of biological diversity is genetic diversity. This is well illustrated by the natural adaptations of brown trout to the rather different environments in which they occur. The Reverend William Houghton's well-known and beautifully illustrated work *British Fresh-Water*

*Fishes* of 1879 lists twelve 'species' of trout. These are the salmon trout, sewen, bull trout, Galway sea trout, short-headed Salmon, silvery salmon, common trout, clack-finned trout, Loch Stennis trout, Loch Leven trout, gillaroo trout and great lake trout (each given a distinct Latin name and an accompanying painting by the artist Alexander Francis Lydon). Current scientific understanding recognises all these as variants of the single species *Salmo trutta*. Of all vertebrates studied to date, the brown trout shows the greatest degree of local genetic structuring. Houghton also describes six species of charr – Windermere charr, Cole's charr, Gray's charr, torgoch (or Welsh charr), Alpine charr and Loch Killin charr – all generally recognised today as the single species *Salvelinus alpinus* (though Torgoch, in particular, has its champions in Wales regarding its unique status as a sub-species). Houghton also recognises four whitefish species – vendace, gwyniad, pollan and powan – the latter three fishes generally regarded today as fragmented populations of *Coregonus lavaretus* (although again there are champions of their distinct status as sub-species or even species). Amongst the glacial relic species – those left behind as fragmented populations in deep, cool and well-oxygenated lakes as the glaciers retreated at the end of the last Ice Age around 11,700 years ago – there will certainly have been natural selection within stranded populations, adapting them over many generations to their unique local situations. So too, brown trout have adapted to the unique characteristics and challenges in different localities to produce a diversity of strains and, over time, natural selection may favour discrete genetic traits. In fact,

Windermere charr, Cole's Charr and Gray's charr, as illustrated by A F Lydon in the Reverend William Houghton's 1879 book *British Fresh-Water Fishes*.

many species of fish and other organisms display variations across their broad global ranges. So, what then is a species, or a sub-species, and where do we draw the boundaries?

Clumping all fishes with similar traits into single uniform species across their often broad geographic ranges, particularly where there are significant environmental barriers between different populations, can work against protection of unique and diverse genetic diversity. For example, large and/or frequent stockings with farm-produced brown trout, roach, barbel or other species can swamp locally adapted strains. Loss of genetic diversity, homogenising the stock and replacing local adaptations and distinctiveness, is as much of a conservation problem as threats to species themselves. This form of 'extinction' of genetic traits may prove serious in the long term as genetic variability, adaptability to changing environmental conditions and overall resilience is lost. Measures such as stocking only with sterile triploid rainbow trout and brown trout are advocated in the Environment Agency's 2003 *National Trout and Grayling Fisheries Strategy* to help avert dilution of localised genetic strains and are now becoming increasingly established standard practices. However, considerable losses of local genetic diversity has already occurred. The genetic heritage of our world is an insurance entrusted to us by billions of years of evolution. We would do well to take rather better care of it than we have to date.

Another important element of biodiversity is age structure within populations. Naturally, age classes within species form a 'pyramid': millions of eggs, producing thousands of fry, hundreds of yearlings and an increasingly diminishing number of older year classes as individuals fall prey to predators, diseases and other natural hazards. A river or still water with poor recruitment (the addition of new individuals to the population) may be attractive to some angling interests when it artificially creates a preponderance of large specimens, but can be an indicator of a fishery that is on the edge of collapse, as has been seen on a number of rivers suffering from environmental pressures.

**Healthy fish populations comprise fish of mixed ages.**

## 11.2 THREATS TO BRITISH FRESHWATER FISHES

The waters in which Britain's freshwater fishes swim are subject to multiple pressures from human development, all of which pose threats to the fishes that live in them. These threats relate to the quality and quantity of water, physical habitat modifications, introductions of inappropriate species, the cumulative jeopardy of a changing climate and a growing human population and their combined impacts.

### 11.2.1 Water quality

The chemical composition of Britain's fresh waters varies naturally. Water running off hard, largely insoluble rock tends to contain lower concentrations of dissolved and suspended substances, creating nutrient-poor streams and pools. Conversely, water running from and through soluble substrate will tend to contain higher concentrations of dissolved matter, such as calcium and magnesium salts responsible for 'hard' water. Also, in the lowlands, richer soils and accumulated inputs from larger upstream catchments, including recirculation of organic matter and nutrient chemicals within rivers and still waters, results in more chemically rich water bodies. Purer water tends to hold a higher concentration of oxygen, particularly where turbulent flows along steeply sloping channels agitate the water's surface, aiding gaseous exchange. Conversely, lowland still and running waters enriched by organic matter, the decay of which absorbs oxygen, tend to have lower oxygen concentrations. Near the coast, saline water exchanges with fresh, changing the nature of the aquatic environment. This variety of natural factors affecting water quality differentially favours the competitive success of different assemblages of fishes.

Overlain on this natural graduation is the influence of human activities. This includes inputs of pollutants from sewage and industrial treatment and run-off from farming, urban and transport

Water running off insoluble rocks tends to have lower concentrations of dissolved and suspended substances and a naturally less profuse assemblage of biodiversity.

systems. Severe pollution can result from excessive inputs from sources such as poorly treated sewage, industrial waste or leakage from cattle slurry stores. Far from all pollutants enter our fresh waters from discrete and generally more controllable 'point sources'. Many arise from 'diffuse' inputs from rural and urban land uses. Substances such as organic matter that tend to strip oxygen from the water as it is broken down by microorganisms and chemical nutrients such as phosphorus and nitrogen that are naturally present at low concentrations, as well as pesticides and other agrochemicals, oil and silt, can all adversely affect freshwater life. Control of both 'point' and 'diffuse' pollutant sources is vital to protect the quality of water upon which fishes, their food and the wider aquatic ecosystem depend.

Some pollutant substances have exotic effects. For example, many synthetic chemicals have endocrine-disrupting effects. What this means is that these substances, such as oestradiol (synthetic oestrogen found in birth control pills) as well as a diversity of surfactants, pesticides and other widespread chemicals, can mimic or block the action of natural hormones. Feminisation of fishes is a widespread problem arising from endocrine-disrupting chemicals across Britain. Other substances entering the water in excess can have largely physical impacts, such as excessive sediment inputs substantially arising from poor agricultural practices. Sediment inputs liberate topsoil into rivers, blocking river gravels essential for successful spawning and early life stages of gravel-spawning fishes such as Atlantic salmon and brown trout as well as barbel, chub and dace. When otherwise suitable gravels become blinded, the fish may fail to be able to cut redds or the eggs falling into spaces in the gravel may not be adequately flushed by clean water, resulting in them asphyxiating.

Deep, naturally clean still waters with naturally high oxygen concentrations at depth are the home of glacial relic species such as Arctic charr, vendace and European whitefish, as well as various other salmonid species. However, rising inputs of nutrient and organic substances can progressively reduce oxygen availability in deeper layers, resulting in the decline of these vulnerable fish populations. In some cases, such as Llyn Padarn in north Wales where the torgoch (Arctic charr) is naturally found, point source inputs from sewage treatment are implicated in the decline of the fish population. More generally, farming and other diffuse activities across wider catchments tend to make the biggest contribution to rising pollutant loads affecting these waters.

### 11.2.2 Water quantity

In some catchments, impermeable geology means that little water is stored underground. Consequently, rainfall results in rapidly rising spate flows, with water levels then declining as rapidly. In catchments on permeable strata, particularly chalk, limestone and sandstone, water tends to percolate into the ground with the result that stream and river flows are buffered, responding only slowly to precipitation, with flows maintained during dry weather. Another side effect of the slow pathway by which water moves through permeable geology is that it produces a more uniform, buffered temperature regime as water temperature is modified by its transit through underground strata. The input of water from underground is known as 'baseflow'. Water reaching rivers and standing waters by overground flow is known as 'runoff'. The ratio of long-term inputs of water entering rivers via groundwater to total stream flow is known as the 'baseflow index', or BFI. A higher BFI is found in rivers running off permeable strata, such as chalk. Stillwater bodies follow a similar pattern, those on hard geologies generally variable in level whereas levels in many water bodies on permeable geologies or else in permeable soils overlaying bedrock in lowland regions respond in a more buffered way to surrounding groundwater levels and flows. Water quantity is, of course, further influenced by the amount and timing of rainfall. Much of the rainfall arriving in Britain is blown in on prevailing southwesterly oceanic winds and falls disproportionately in the west, with a 'rain shadow' in the south and east which is exactly where the human population is densest and demands on water resources are highest.

The highly permeable nature of chalk geology results in rivers with a high baseflow index and, hence, buffered flow and temperature regimes.

Abstraction of water from rivers, lakes and reservoirs can radically affect the quantity of water remaining in them and the flow of running waters. More pernicious is extraction from groundwater, offering economic benefits as this source is generally purer, requiring less treatment with associated costs before use directly or in public supply networks. However, excessive extraction of groundwater can deplete resources, contributing to river flows and the level of connected still waters. Declining baseflow can radically alter flow regimes in rivers, geomorphology (habitat-forming processes), suitability for different species and susceptibility to drought or other extended dry periods. Declining flows in rivers also enable saline water to push further up estuaries and to infiltrate into coastal groundwater. Saline intrusion squeezes freshwater habitats, as well as potentially compromising transitional water habitats available to species such as European eels, Atlantic salmon and twaite shad that require a period of adjustment when moving between fresh and saline water and that support species commonly found in estuaries such as flounder, smelt and sand smelt.

Creation of large areas of impermeable surfaces through development activities can also exert significant effects on water availability by preventing the natural infiltration of moisture downwards into soils and the replenishment of groundwater. Paved surfaces, roads and roofs all have obvious effects, but perhaps more widespread is the 'panning' of soil surfaces due to heavy farm machinery or trampling by livestock at high densities that can compress the soil surface, forming an impenetrable layer. Aside from a reduced rate of groundwater replenishment, impermeable surfaces also result in surges of runoff from wide areas after rainfall, increasing the 'flashiness' of rivers and pulses of water entering lakes, ditches, canals and estuaries, and the likelihood of carrying impurities picked up en route, such as eroded surface soil, oils, soil, grit, rubber and other vehicle breakdown products, amongst a range of other substances.

### 11.2.3 Habitat degradation

Another widespread impact arising from the pressures we put on our rivers and landscapes is the loss or degradation of habitat in and adjacent to flowing and still waters, both large and small, as well as surrounding estuaries. Such is the pressurised nature of contemporary landscapes that paucity of necessary habitat can seriously constrain freshwater ecosystems, including their fish populations. Declining, or conversely protected or rehabilitated, habitat can also make major contributions to the resilience of ecosystems to other compound pressures, such as those arising from declining or variable water quality and water levels. As we have seen, an abundance and diversity of appropriate habitat is essential for the success of freshwater fish populations, supporting a mixture of species and the full range of life stages from spawning and fry nursery to adulthood, providing for their feeding, refuge and reproductive needs.

Estuaries are often highly encroached by development of industrial, urban, residential, port, resort and other infrastructure. (Image © Dr Mark Everard.)

However, over-management of water bodies, both directly and in catchments affecting them, is a pervasive problem. From direct river, still water and drainage channel bank engineering to weirs that block the free movement of fishes and sediment, land drainage for agricultural, urban and industrial and infrastructure uses, today's freshwater environments are barely recognisable from their truly 'natural' state. The substantial modification of the British landscape and waterscape, profoundly changed today from its original character, has been discussed at length in the chapter considering 'Freshwater fish ecosystems'. Since Roman times, Britain has lost an estimated 90% of its wetlands. Not only that, but our remaining, much simplified fresh water bodies have lost the connectivity with surrounding landscapes, successional wetland systems and the wider workings of the water cycle. A survey in 2003 across England and Wales found that only 7.1% of river stretches had a floodplain on either bank for at least a third of their length, and 13% had artificial embankments close to the watercourse and 6% had embankments set a little further back. A subsequent 2010 survey found that 42% of surveyed river stretches were 'severely modified' in terms of habitat, with a further 20% 'significantly modified'. These summary figures indicate the extent of physical impact on the rivers of the British Isles. Another survey of rivers across England and Wales concluded that 42% of former floodplain has been lost. We drain, till and graze riparian lands, often right up to the water's edge, build our cities like strait jackets constraining the very life-giving flows that one provided the water resource, fishery, wildfowling, navigation, defensive and other benefits upon which they were founded. Bridges further constrain high flows, and port and resort development radically squeeze transitional waters at estuaries.

Habitat loss, from the ploughing over of small ponds to increase field size, urban and industrial sprawl, infrastructure and other conversions of landscapes, radically changes the form and function of Britain's diverse waterscapes and, with it, prospects for our freshwater fish fauna.

### 11.2.4 Introduced species

A wide diversity of organisms have been introduced beyond their natural range into Britain, as indeed around the world. As this book is about freshwater fishes, we will address introduced fish species separately in the following section. This section addresses the impacts of other types of organisms on fishes and the ecosystems of which they are part.

Whilst many introduced species prosper without adverse consequence, a wide range of plants and animals freed of the constraints of the ecosystems with which they co-evolved can explode in numbers or otherwise exert a destabilising effect on the new ecosystems within which they become established. Once these genes, like genies, are out of their metaphorical bottle, unpredictable, often detrimental and sometimes irreversible consequences can occur. The reality is that our rivers and fresh still waters are under attack by many alien species that were not co-evolved within them, perturbing or with the potential to disturb their fine balances. Many instances of this are observed right around the world. 'Genetic homogenisation' is a term sometimes used to describe the tendency of invasive species to overtake locally adapted ecosystems, outcompeting or expunging native genetic stock with the result that the gene pool of ecosystems tends to become simplified and common over wide areas. The invasion of freshwater ecosystems by common carp, profoundly changing the quality and diversity of the ecosystems in which they proliferate and contributing to the decline of native aquatic vegetation, fish and other species, is one commonplace example.

Many plants have been translocated around the world for ornamental and horticultural purposes, some of them now outcompeting native species. This can have adverse impacts both directly for the stability of soils and the wider ecosystems in which they grow, but perhaps more importantly by breaking linkages in the life cycles of insects and other organisms dependent upon native vegetative diversity. Some of these invasive plants are aquatic, as, for example, Canadian pondweed (*Elodea canadensis*), native to most of North America, but since introduced widely around the world and first recorded from the British Isles around 1836. Canadian pondweed is now so pervasive in still, fast and slow-flowing fresh waters across Britain that it might be assumed to be native. However, in the 1960s, its rapid spread through canals and other bodies led to huge growths, swamping native plant species and reportedly stranding some fish. In many waters, Canadian pondweed can form dense beds covering virtually the whole bed of a pool or river margin where there is sufficient light. Though growing perennially, substantial die-back in autumn and winter can result in deoxygenation of lower layers of water as large volumes of plant matter decay. A more recent appearance in British fresh waters is the swamp stonecrop

*Crassula helmsii*, swamp stonecrop, is a problematic invasive plant growing from tiny propagules, now banned from sale but only after widespread invasion of British freshwater bodies. (Image © Dr Mark Everard.)

(*Crassula helmsii*), also known as New Zealand pigmyweed or Australian swamp stonecrop, first recorded in Britain in the New Forest in 1976 in a roadside pond adjacent to houses and spreading to 194 ponds and ditches in the New Forest by 2000. This tiny but prolific plant has spread widely throughout Britain by transfer, but also wider sale in garden centres. Swamp stonecrop poses a significant and increasing risk to freshwater habitats, as it can form dense clumps as both a submerged and an emergent plant, swamping other native water plants and the invertebrates they support. It can grow from tiny, hardy propagules that are easily transferred from water to water. The sale of swamp stonecrop is now banned across Britain, but arguably way too late to make a significant difference to its potential to cause harm. Many other introduced aquatic plants are also now problematic, with legal controls generally only put in place many years after severe adverse impacts have been observed and arguably too late to prevent further invasion.

However, it is not just plants in the water that pose threats for the habitats of freshwater fishes and other aquatic life. Bankside vegetation too can displace native species, perturbing

**Himalayan balsam can outcompete native vegetation, excluding the insect and other life that it supports, also dying back in winter to expose soils to riparian erosion.**

ecosystems. Japanese knotweed (*Fallopia japonica*), for example, can form dense stands on river and lake banks as well as roadsides. It is easily spread, swamping native vegetation and providing little useful food or habitat for other wildlife. Another plant that is attractive but a significant cause for concern for river structure and life is Himalayan balsam (*Impatiens glandulifera*), introduced through horticulture in 1839 from its native range in the foothills of the Indian and Pakistani Himalayas and spreading, in the absence of natural enemies to keep it in check, to rapidly become one of the UK's most widespread invasive weed species. Once established, Himalayan balsam colonises river and still-water banks as well as numerous other damp habitats, spreading rapidly as tall stands. It outcompetes native plant species for space, light, nutrients and pollinators, so reducing native biodiversity. Further problems arise as Himalayan balsam is a shallow-rooted annual, dying back in the winter to expose bare ground that is highly susceptible to erosion. This brings silt, nutrient chemicals and other associated substances into water bodies, where they blind spawning gravels, degrade riparian habitat structure and otherwise adversely affect fish and other aquatic wildlife.

In the average garden centre, rows and rows of non-native plants and genetic strains are available on open sale. The British flora is massively influenced by escapes from horticulture, though the popularity of gardening and the economic might of the long-established trade does rather blind people to potentially severe risks of vegetative invasion. Only after many years of demonstrable adverse impacts does trade in plants such as Himalayan balsam, giant hogweed (*Heracleum mantegazzianum*), Japanese knotweed, floating pennywort (*Hydrocotyle ranunculoides*) or Australian swamp stonecrop become restricted and their occurrence notifiable. By then, the genies are out of the bottle, and effective control may be impossible.

Potentially problematic introduced animals include various crayfish species, in particular, the large, voracious and aggressive American signal crayfish (*Pacifastacus leniusculus*). Together with a number of other invasive alien crayfish species such as the Turkish crayfish (*Astacus leptodactylus*), this alien crustacean not only outcompetes the sole native crayfish species, the white-tipped crayfish (*Austropotamobius pallipes*), but also spreads the spores of fungal disease that is 100% lethal to the native crustaceans. The white-tipped crayfish is now severely threatened. Meanwhile, the American signal crayfish continues to spread, often seemingly by illegal introductions by people who then harvest established populations for lucrative trade serving the catering industry. Whilst the white-tipped crayfish is smaller and co-evolved with British ecosystems, the American signal crayfish is a voracious feeder on water plants, fish eggs and larvae, as well as other invertebrates and also tends to burrow into banks, in some cases leading to the collapse of weirs and banks imposing significant costs and to the great detriment of freshwater fishes and other aquatic wildlife.

Many more invasive animals are found in British fresh waters, including, for example, two species of 'killer shrimp' (*Dikerogammarus villosus* and *Dikerogammarus haemobaphes*), native to the Ponto-Caspian Region of Eastern Europe. These killer shrimp species were first found in Britain, respectively, in 2010 and 2012. They consume a diversity of invertebrates

The signal crayfish is a large, invasive crayfish originally from North America, causing direct harm in British fresh waters but also carrying a fungal disease that is 100% fatal to the native white-clawed crayfish.

**The 'killer shrimp',** *Dikerogammarus villosus*, **first appeared in British fresh waters in 2010 and poses significant threats to aquatic ecosystems. (Image credit: Environment Agency.)**

species as well as young fish and so can significantly alter ecosystems. The zebra mussel (*Dreissena polymorpha*), once found only in the lakes of southeast Russia, has become a near-global pest that forms dense colonies, smothering the beds and banks of fresh waters, and it can also clog water pipes.

What is overlooked here is the potential for spread of microscopic organisms. This is largely because they escape our scrutiny. Yet we know that bacterial, fungi, archaea and a range of other microscopic organisms exert profound impacts on the functioning of ecosystems from photosynthesis to nitrogen cycling and many more processes foundational to healthy ecosystems. One such microscopic plant that has been attracting attention is the delightfully named 'Rock Snot' (*Didymosphenia geminate*), also more politely known as Didymo, which is a species of diatom (a type of alga) that produces nuisance growths in freshwater rivers and streams with consistently cold water temperatures and low nutrient levels. The species seemingly originates from northern America, but is considered invasive in Australia, Argentina, New Zealand and Chile, as well as problematic within its native range where it has taken on invasive characteristics since the 1980s. Though not considered a human health risk, it can make angling and other activities unpleasant but, more significantly, can affect stream habitats and sources of food for fishes and other aquatic organisms.

The simple lesson is that species introduced beyond the native range in which they co-evolved can be profoundly problematic and also potentially impossible to eradicate once established. Prevention then is better than cure, in terms of bringing new species into the country as well as spreading both introduced and native species beyond their adapted natural range.

### 11.2.5 Threats from aquaculture

The topic of aquaculture has been touched upon variously throughout this book, initially in the context of early Roman and British monastic fish ponds. It will also be touched upon subsequently as we turn our considerations to stocking of fisheries, as well as in other aspects of conservation. Globally, fish production through aquaculture is growing rapidly in scale and is likely to exceed fish landings from wild capture fisheries by 2030. Aquaculture has the potential to be a sustainable source of fish and other seafood, contributing to alleviating the significant and often overbearing pressures that we place on the world's oceans and other aquatic resources. However, like any intensive farming system driven by profit maximisation, often in ignorance or disregard for wider implications, the sustainability of current stewardship of fish farms is highly questionable because of organic, nutrient, pharmaceutical and other pollutants escaping the farms. Further problems arise from parasite and disease transmission, farmed fish escapes, the nature of the feed given to them and a range of other problematic issues. The scale and means of aquaculture production pose often significant threats to British freshwater fishes.

The acceptability and market for cyprinid and other groups of freshwater fishes other than salmon and trout is low in Britain, somewhat at odds with much of the rest of Europe. As a consequence, the vast bulk of fish production in Britain, as also Denmark, Norway, Canada and Chile, comprises salmonid fishes (species from the salmon family) produced both for human food and stocking of waters for recreational fishing.

Both brown trout and the hardier, faster-growing rainbow trout are now extensively cultured. Brown trout have been farmed at low intensity in Britain since the late nineteenth century, largely using enclosed pond systems but giving rise to water quality, disease, escapee and other problems. However, commercial-scale trout farming using elongated through-flow ponds, fed upstream by river or spring water and draining downstream into watercourses, was introduced from Denmark in the 1950s and had grown in the first decade of the twenty-first century to some 360 trout farms across the UK. The introduction of the hardier, faster-growing rainbow trout, native to the Pacific seaboard of America but now widely introduced globally,

**Farming of trout, here being fed in a through-flow channel, can be efficient means to produce food, but also may have a range of associated problems impacting native freshwater fish populations. (Image © Dr Mark Everard.)**

accelerated the pace of trout production substantially. However, all of this does not come without risks for the river systems in which trout farms occur. Pollution in spent water exiting fish farms can cause significant problems related to organic and nutrient chemicals, silt and veterinary medicines, though such discharges have to be licensed and controlled. Fish farms also tend to entrap the fry of other fish species drawn into the fish ponds. They also attract piscivorous predators that can then feed on the neighbouring river and lake systems and represent sources of diseases and parasites released into neighbouring watercourses.

The rapid expansion of caged Atlantic salmon farming is posing significant problems, particularly clustered around the bays and inlets of the west coast of Scotland. A wide range of adverse impacts from modern salmon farming practices includes nutrient enrichment and habitat alteration. Release of parasites into adjacent waters, particularly the spread of sea lice (*Lepeophtheirus salmonis*), is a serious cause for concern as substantially increased infestation of wild sea trout and Atlantic salmon – with many farms located on migration routes for these

Caged salmon farms, often positioned near migration routes of wild salmonid fishes, can he highly problematic for wild fish populations. (Image credit: Richardjohnson/Shutterstock.com.)

species – reduces their resistance to infection and compromises the functioning of the gills, particularly with regard to their capacity to 'pump' salts into and out of the blood.

Escapee trout and salmon from fish farms can also interact with wild fish populations, in terms of feeding and breeding behaviour, as well as dilution of locally adapted gene pools (a topic to which we will return later in this book). Further concerns arise from extensive exploitation of wild marine fish species, particularly sand eels, used in the production of fish meal pellets used to feed farmed carnivorous fish, such as trout and salmon. This can not only detrimentally affect harvested fish populations but can also deplete food resources available for other fishes and for coastal birds. Arctic terns, puffins, guillemots and razorbills are undergoing significant population declines, substantially attributed to declining food resources.

Aquaculture undoubtedly has a role to play in feeding a growing global population, but solutions such as moving to closed-loop systems that avoid parasite and pollutant releases and the adaptation of, and alternative food for, farmed fish need to be addressed as a priority if wider detrimental outcomes – for fishes, for wildlife and for the functioning of whole ecosystems supporting many dimensions of human wellbeing – are to be minimised or averted.

### 11.2.6 Compounding factors

These pressures do not occur in isolation, but have compounding effects. For example, the physical habitat conversion of a weir or dam may slow down flow substantially, reducing the exchange between water and air and decreasing oxygen concentration. Stiller water can also settle out fine sediment, the organic constituent of which continues to decay further, drawing down oxygen levels in the water. Furthermore, the more uniform flows and muddier bed characteristics may favour the spread of invasive plants, reducing diversity of food as well as spawning places for fishes. Any one factor can influence others, for example, water abstraction from groundwater may reduce flows in rivers and draw down water levels in ponds and lakes as well as canals, ditches and reservoirs, making water bodies more susceptible to drought conditions, reducing dilution of contaminants and preventing the free movement as well as refuge, feeding, breeding and nursery areas of fishes.

All of these pressures are further compounded by a changing climate, including shifting patterns and extremes of rainfall and temperature. Globally, climate change is regarded as one of the major factors accounting for the decline of freshwater fish populations. As temperature and rainfall regimes change – water systems are the main vectors by which increasing energy in the climate system manifests – freshwater fishes and many other organisms respond by shifting their distribution range, altering the timing of migration and spawning and through other processes. We have already reviewed the importance of phenology for the diverse interactions amongst fishes and all other components of complex ecosystems in the chapter dealing with 'The things that Britain's freshwater fishes eat'. As mean daily rainfall changes across the year, other species respond in different ways to changing rainfall patterns and temperatures, and the demands of society on farmland and water resources shift in a changed environment, freshwater fish will undoubtedly be subject to mounting pressures.

Fish are therefore under great and compounding pressures from multiple sources, including the diverse demands of an increasing human population, more intensive farming and urbanisation. So fish communities are perhaps one of the strongest indicators of how our fresh water environments are faring.

### 11.3 PROBLEMS ASSOCIATED WITH INTRODUCED FISHES

Freshwater fishes have been widely introduced beyond their native ranges, often multiple times and for a range of reasons, and consequently may be the most manipulated of all elements of the British fauna. Whilst some sectors of society may enjoy non-native fish species for ornamental, aquaculture or sporting purposes, their impacts on our native ecosystems may be potentially far from benign.

### 11.3.1 Reasons for fish introductions beyond their native range

We have seen how the common carp has been repeatedly introduced to these islands from the fifteenth century onwards. Initially, as has been the case nearly globally, common carp were introduced as a hardy, fast-growing source of food. However, subsequently, introductions have been associated with their use as an ornamental species and, more latterly, as a prized recreational angling target.

Barbel are a fish species native to Britain, so perhaps not ostensibly 'introduced'. However, due to Great Britain's former land connection with continental Europe up to the last Ice Age, barbel are native only to drainage basins from what is now the Humber in the north to the Thames basin in the south. They have subsequently been widely spread throughout rivers and even into some still waters across mainland Britain, mainly by angling interests. Rainbow trout are another species introduced here for both aquaculture and sporting purposes, and limited gene pools of the otherwise native brown trout have been widely stocked into waters across Britain, diluting locally adapted strains. These trout species have now also been introduced widely across much of the world where conditions are conducive.

Other non-native species, such as topmouth gudgeon, sunbleak, goldfish and orfe, have escaped, or been released, from aquaria and ponds, proceeding to establish wild populations. Some other species have been spread, in part, by accident, such as the appearance of ruffe in the Kennet and Avon Canal as 'stowaways' in stockings of other fish already found in the canal. Many introduced species, such as common carp, wels catfish, rainbow trout and zander, have been here for a long time and are now more generally considered part of the British fish fauna despite the considerable harm that some of them can cause when spread out of place.

**The barbel is native to rivers draining to the east of England from the Thames system in the south to the Humber catchment in the north, though it has been widely introduced across Britain.**

### 11.3.2 Invasive fish species

Not all introduced freshwater fish species are problematic. Take, for example, the bitterling, naturalised in a few scattered localities across England probably since their popularity in cold water aquaria from the end of the nineteenth century. These small fish do not seem to have perturbed ecosystems to any notable extent. Others, however, can become distinctly invasive and problematic.

An invasive species is any organism that is not native and which has a tendency to spread to a degree believed to cause damage to the environment, economy or human health. Some freshwater fish species, including topmouth gudgeon, sunbleak, bitterling, pumpkinseed sunfish and black bullhead, share some or all of the attributes generally considered to predispose aquatic organisms to become invasive, summarised in Box 11.1.

I have already referred to common carp as 'pigs with fins', the analogy being precise in many regards as the species is hardy, breeds readily, eats almost anything and grows quickly, providing a useful protein source. It also grubs up the sediment, devastating the ecosystem if present in dense populations. Although now widely introduced almost pan-globally, common carp are native to the Caspian Sea region, migrating naturally to the Black and Aral Seas and their tributary rivers and then eastwards to eastern mainland Asia. There is debate about whether the River Danube was part of the native range of this species. However, what is beyond doubt is that common carp are now naturalised widely across Europe, Asia, Australasia, Africa and the Americas owing to introductions by people. Both Romans and monks were responsible for the spread of common carp in Europe, selectively breeding useful strains particularly for food. As common carp serve this pig-like role,

## BOX 11.1 ATTRIBUTES OF AQUATIC ORGANISMS PREDISPOSED TO BECOME INVASIVE

The following attributes were recognised by Ricciardi and Rasmussen (1998) as predisposing aquatic organisms to becoming invasive:

1. Abundant and widely distributed in their original range
2. Wide environmental tolerance
3. High genetic variability
4. Short generation time
5. Rapid growth
6. Early sexual maturity
7. High reproductive capacity
8. Broad diet (opportunistic feeding)
9. Gregariousness
10. Possessing natural mechanisms of rapid dispersal
11. Commensal with human activity (e.g. transport in ship ballast water or trade of ornamental species for aquarists)

latterly also with ornamental and angling uses, the species has effectively 'hitchhiked' on humans to populate much of the temperate and tropical worlds. Here, it has outcompeted other species or, because of feeding methods, grubbed up sediment to turn clear water, vegetated ecosystems into murky waters that are more expensive to clean up for other uses as well as impoverished in native biodiversity. As a scientist, I have worked on freshwater systems – from Australia's Murray-Darling river system to formerly biodiverse, clear-water pools in Britain, Kenya, South Africa, India and China and across to the US – where rapid population growth after introduction of common carp has radically changed the ecology and quality of fresh waters, including contributing to dangerous algal blooms and the extinction of native fish species. Common carp in their native range are co-evolved as integral elements of ecosystems; out of these places, they can be hugely destructive with many ramifications for people. In my eyes, common carp out of place will ever remain unwelcome 'pigs with fins'.

The adverse impacts of invasive species differ. For example, the goldfish has been widely introduced from the ornamental fish trade, surviving and prospering in our rivers and still waters. This has been detrimental specifically to the closely related, native and increasingly endangered crucian carp with which it hybridises. So widespread are goldfish in their native brown form that it is hard to determine how many 'pure' genetic strains of crucian carp now remain.

Another species that can be considered invasive, despite being a native freshwater fish, is the ruffe. Ruffe are naturally residents of the slacker waters to the east and south of England. Though they are not a primary target for anglers, they have sometimes been used as bait for predatory species, and it is by this route that some of their more damaging introductions

**Common carp, 'pigs with fins', have been widely introduced across the world as they are efficient in turning many food sources rapidly into animal protein, much as pigs do on land.**

**Introduced ruffe are implicated in the extinction of vendace from Bassenthwaite in the English Lake District.**

appear to have occurred in more northerly British waters where they were formerly absent. Bassenthwaite in the English Lake District once held the only English population of vendace, a rare whitefish. However, introduction and the subsequent population explosion of Ruffe from the 1980s onwards has, allied with rising diffuse inputs of nutrients to the lake, made a major contribution to the eventual extinction of vendace from the lake. Ruffe predate the eggs and fry of vendace and other species, constituting a major pressure amongst a range of other environmental problems that have now wiped out the sole English population of vendace. To the north in lowland Scotland, ruffe have also been introduced into and spread within Loch Lomond and are implicated as a major factor amongst a range of environmental issues in the decline in the population of powan (the local name and strain of European whitefish). Further afield, European ruffe have also been accidentally introduced to the Great Lakes system of the US, most probably from ballast water, where they have also become a highly significant pest.

Another invasive freshwater fish species making a significant nuisance of itself in Britain is the topmouth gudgeon, introduced through the ornamental fish trade from its native east-Asian range into various areas of Europe and Asia. These small, resilient fish are adaptable, can breed after only their first year of life and are thereafter capable of spawning several times per year, even in poor conditions. Although topmouth gudgeon produce relatively few eggs in each batch, adhering them to stones or leaves throughout the spring and summer months, male fish then subsequently guard them and so fry survival is greatly enhanced. Topmouth gudgeon may thereby rapidly colonise a freshwater body following introduction, quickly outcompeting other fish, then proceeding to eat their eggs and juveniles, effectively preventing recruitment of other species. This threat is more than theoretical and has manifested on a range of British waters from Cumbria southwards to Devon, often necessitating drastic eradication campaigns using chemical poisons.

The topmouth gudgeon is not the only problematic invasive freshwater fish species exerting this effect on native fish populations. The sunbleak, native to areas of Asia and eastern Europe, has also spread across Northern Europe through releases from the aquatic trade and via canal systems. Sunbleak also rapidly grow to adult size, can breed after their first year and thereafter spawn multiple times each year. Male sunbleak also guard eggs attached to water plants, enhancing their survival. Sunbleak can thereby rapidly colonise any fresh water after introduction, which may be through eggs adhering to water plants or nets and can have the same impact as topmouth gudgeon. However, whereas topmouth gudgeon are fishes of still and slow-moving water, sunbleak also populate rivers and so may pose a greater threat in the long term as eradication from non-enclosed waters is all but impossible.

The spread of other species, introduced both from overseas or beyond their native range, can have further detrimental impacts. This includes perturbation of locally adapted ecosystems, but also spread of associated parasites. By and large, fish introductions present major risks from an ecological perspective and have to be considered carefully and with a great deal of precaution and compliance with regulatory restrictions.

### 11.3.3 Prevention is better than cure

If the control of introduction of plant species is retrospective, often occurring years or decades after widespread negative impacts have been demonstrated (Himalayan balsam, giant hogweed, Japanese knotweed, floating pennywort and swamp stonecrop are some examples of such species now banned from sale), the control of potentially alien invasive fish species is now rather better considered, at least for more recent invasive species and those that might become so.

Threats posed from foreign freshwater fish species to the balance and integrity of British freshwater ecosystems are addressed legally on the basis of their potential to escape and form self-perpetuating populations. The Import of Live Fish (England and Wales) Act 1980, or ILFA for short, is the primary piece of legislation enacting this. In 1998, the Prohibition of Keeping or Release of Live Fish (Specified Species) Order was also introduced in an effort to reduce the number of fish species introduced into fresh waters in England and Wales, adding further species to the list of those initially scheduled under ILFA. These legal instruments collectively schedule temperate freshwater fish species believed likely to threaten the environment owing to their potential for forming self-sustaining populations, with unpredictable outcomes including perturbation of the balance of native aquatic ecosystems. The legislation stipulates that suppliers need a licence to sell listed species, and fish keepers also require a special licence to keep some of the scheduled fishes. The Prohibition of Keeping or Release (Specified Species) Scotland Order 2003 is largely similar to the Order for England and Wales. A list of scheduled freshwater fish species is included in Appendix 5.

## 11.4 NATURE CONSERVATION AND BRITAIN'S FRESHWATER FISHES

Whilst some nature conservation and other legislative initiatives have, as we will see, been explicitly aimed at selected fish species, many more have been informed by the needs of freshwater fish species. In this section, we will look at a range of legislative and policy initiatives that, whether directly or indirectly, respond to the needs of Britain's freshwater fishes.

### 11.4.1 History of conservation affecting Britain's fresh waters and their fishes

Regulation of river quality in the UK has progressed through multiple phases. Up to the 1970s, the primary focus was on sources of pollution, with a particular emphasis on industrial discharges and sewage treatment works. From the 1970s onwards, emphasis shifted to a set of 'use-related' standards relating to the chemical quality of the receiving waters, significantly informed by the needs of the types of fish populations expected to occur in them. These use-related standards for classifying and managing discharges to water bodies was superseded in 2000 by the new regulatory focus of the EU Water Framework Directive (WFD), the end point of which rested not on the source of quality deterioration nor on identifying uses of the receiving water but instead on the achievement of 'Good Ecological Quality' (or 'Good Ecological Potential' where overriding economic factors prevent achievement of Good Ecological Quality) of designated Water Bodies assessed across a range of chemical, biological and hydromorphological parameters. The WFD extends to fresh surface and groundwater, inshore and transitional waters. Under the WFD, fish populations become an ecological end point, rather than simply a basis for setting chemical standards. There, of course, remain simplifications in how ecological status is assessed, legislative approaches generally being necessarily coarse in order to be affordable and otherwise practicable.

Though the substantial body of contemporary legislation pertaining to protecting freshwater and other ecosystems is without precedent, there is a perception amongst many in the public arena of an 'implementation gap'. Holistic regard for rivers and other fresh waters, and adverse trends in the quality of certain taxa particularly of migratory fishes integrating pressures across whole catchments in completion of their life cycles was still somehow being missed. This was the spur for genesis of establishment of civil society Rivers Trusts across the UK from the late 1980s onwards. By the 2010s, the UK's Rivers Trusts movement subsequently evolved and spread across the country, proving influential in integrating the efforts of existing statutory bodies and non-statutory specialist interest and community-based local catchment management groups. The niche filled by the UK's Rivers Trusts is one of taking an holistic overview of problems with rivers, informed by local knowledge and context and achieved by public participation, collaborative working and decentralised modes of assessment, planning and decision-making. This engaged approach has not only improved progress with implementation of the WFD, but also filled a democratic gap in transposition of the Directive into the UK that had formerly overlooked the requirement for public participation. This in turn has led to co-learning between statutory and non-government actors, the successes of the participatory approach directly informing the English Government's Catchment-Based Approach (CaBA) policy to relocate debate, decision-making and action about the future direction of improvements to the water environment more locally. The CaBA embeds collaborative working at a river catchment scale to deliver cross-cutting improvements to the water environment and is active in the 100+ WFD catchments across England, including those cross-border with Wales. CaBA engages with more than 1,500 organisations, addressing multiple benefits ranging from flood risk to water quality, biodiversity and climate change.

The evolution of river quality management, with the health of Britain's freshwater fishes very much to the fore, thus represents a successful model of transition from problem characterisation and reaction towards a stakeholder-based mission to address locally identified needs spanning multiple, formerly fragmented disciplines. Fish also are seen not merely as a basis for the setting of chemical standards, but ultimately as the indicators, along with other aquatic animals and plants, of the vitality of running, standing and transitional waters.

As a corollary to this generally positive story, the UK government set a suite of indicators of sustainable development with associated targets in the late 1990s. These indicators explicitly included the status of Atlantic salmon as both an indicator of ecosystem health and as an iconic species. The Department of Environment, Food and Rural Affairs (Defra) included the 'Number of rivers in England with sustainable salmon stocks' as one of nine headline biodiversity indicators, with the objective of protecting and enhancing Atlantic salmon stocks to ensure sustainable exploitation by fisheries as well as conservation of genetic diversity. The indicator relates to the number of rivers in which Atlantic salmon spawning levels have met

**Conservation Limits for Atlantic salmon were set as a key sustainable development indicator by the UK government, an indicator never met and subsequently abandoned.**

or exceeded agreed Conservation Limit standards for sustaining the native salmon stock. The Conservation Limit is set at a stock size below which further reductions in spawning numbers are likely to result in significant reductions in the numbers of juvenile fish produced in the next generation and, hence, returning adults from that year class. It also recognises that, as each river has a genetically distinct salmon stock, once the population of salmon is reduced below the number required to maintain a sustainable population, there is a significant risk of losing that genetic diversity. A target of twenty-seven qualifying rivers out of forty rivers assessed annually was set in 1999. Between 1997 and 2008, the number of rivers with sustainable salmon stocks varied between thirteen and twenty-eight, with a tentative upward overall trend in the later years, perhaps related to more intense habitat and other directed conservation activities. Though welcome, this gives no leeway for complacency, as forty rivers meeting their Conservation Limit targets out of the limited subset of forty surveyed is surely the only logical target for a sustainable future in which critical aquatic and other environmental resources are not being systematically undermined by the cumulative pressures of contemporary lifestyles. However, the bad news is that the UK government subsequently abandoned this sustainable development indicator, undermining prospects both for fish and the long-term security and wellbeing of British society.

### 11.4.2 Freshwater fish-related nature conservation agreements and legislation

Many freshwater fish species are of direct nature conservation concern, key parts of the biodiversity of freshwater habitats that are regarded as amongst the most threatened ecosystems globally. Of Europe's freshwater fish species, 38% are threatened with extinction, with a further twelve fish species already declared extinct. At a broader scale, approximately 20% of the world's 15,000 freshwater fish species are listed as threatened, endangered or extinct. Our concern for fish should be far more than altruistic, as this decline indicates a commensurate reduction in the vitality of ecosystems, including their capacities to support our various needs into the future.

Having noted the more general lack of integration of Britain's freshwater fishes into wider consideration of 'wildlife', a number of them are explicitly recognised in legislation as of nature conservation concern. For some, such as Arctic charr and Atlantic salmon, declining populations demand management action. Other non-native species, such as topmouth gudgeon and black bullhead, pose a threat to native species and ecosystems. Their introduction to the fresh waters of the British Isles or their further spread therefore needs to be strictly controlled. This is essential as, in addition to their roles in cycling food, energy and parasites, freshwater fishes are an important part of freshwater ecosystems at all levels in the food chain. Their conservation or, in the case of problematic species, elimination or appropriate management, is vital for the future of healthy and sustainably managed watercourses and other wetland habitats.

Key international agreements and pieces of legislation scheduling selected British freshwater fish species for special protection based on their nature conservation importance include:

- The 'Red List', or more properly the International Union for Conservation of Nature (IUCN) 'Red List of Threatened Species', is a comprehensive inventory of the global conservation status of species of plants and animals. The Red List is regularly updated by expert committees.
- The Bern Convention (or Berne Convention), which has the full title 'The Bern Convention on the Conservation of European Wildlife and Natural Habitats 1979', schedules species of plants and animals threatened across their European range. The Bern Convention has four principal aims: (1) conserve wild flora and fauna and their natural habitats, (2) promote cooperation between states, (3) monitor and control endangered and vulnerable species and (4) help with the provision of assistance concerning legal and scientific issues.
- The European Union (EU) Habitats Directive was adopted in 1992 as a cornerstone, jointly with the EU Birds Directive, of Europe's nature conservation policy. The EU Habitats Directive has been variously amended since adoption. It has the two central pillars of designation of a 'Natura 2000' network of protected sites and a strict system of species protection. Overall, the EU Habitats Directive covers a wide range of animal and plant species and over 200 'habitat types' of European importance and is the means by which the EU meets obligations under the Bern Convention.
- At the UK level, the Wildlife and Countryside Act 1981, and its many subsequent amendments, implements the EU Birds Directive but also consolidates various other measures protective of native species, especially those under threat. It also places controls on the release of non-native species, enhances protection of Sites of Special Scientific

Interest (SSSIs) and builds upon rights of way rules in the National Parks and Access to the Countryside Act 1949. In England and Wales, the Countryside and Rights of Way (CROW) Act 2000 made new provisions for public access to the countryside and amended the law relating to nature conservation, protection of wildlife and protection of areas of outstanding natural beauty. As the CROW explicitly excluded new measures on rights to fish in non-tidal waters, the provisions of the prior legislation remain largely in effect.

- The UK Biodiversity Action Plan (UK BAP) is the UK Government's response to the international Convention on Biological Diversity (CBD), to which the UK became a signatory at the 1992 United Nations 'Earth Summit' in Rio de Janeiro. The CBD called for the development and enforcement of national strategies and associated action plans to identify, conserve and protect existing biological diversity and to enhance it wherever possible.

Further details concerning these agreements and pieces of legislation are included in Appendix 6, together with their various Schedules and Lists and the British freshwater fish species to which they relate.

### 11.4.3 Fish introductions and the law

The Import of Live Fish (England and Wales) Act 1980, or ILFA, has already been introduced in this chapter. It will not be described again here, save to say that although enacted in response to observed, often severe consequences, it is also a foresighted piece of legislation applying not only to a range of Scheduled species, but also in principal controlling introductions of any species of fish capable of forming self-sustaining populations in British waters. As observed above, prevention is far better than a cure that, in the case of control of plant imports, tends to come way too late for effective eradication or control of problematic invasive species.

### 11.4.4 Fishes of value to other species of conservation concern

As fishes serve many important roles in food chains at all trophic levels, they play an important part in the functioning and resilience of aquatic and associated ecosystems. Some are also an important food source for species of conservation concern. A good example is the rudd, which favours standing and slow-moving waters, some of which also comprise the reed bed fringe habitat favoured by the great bittern, one of Britain's rarest birds that is included in the UK BAP. The European eel, a fish species now listed as Critically Endangered on the IUCN Red List, is also a favoured food for BAP species including the European otter and various birds covered by the EU Birds Directive and the UK's Wildlife and Countryside Act 1981 including the common (or Eurasian) kingfisher (*Alcedo atthis*), great cormorant (*Phalacrocorax carbo*) and grey heron (*Ardea cinerea*).

### 11.4.5 The values of other species of conservation concern to freshwater fishes

Further multi-beneficial outcomes have been observed where large animals have been either reintroduced after they had been driven locally to extinction, or else where their populations

The common kingfisher is an attractive and widespread piscivorous bird, integral to freshwater ecosystems and protected under European and UK legislation.

have been allowed to recover. Larger animals, particularly carnivores, are commonly the most impacted organisms from human-driven habitat conversion, human–animal conflict or persecution. Many of these species can play significant roles as 'ecosystem engineers', changing the structure or functioning of whole ecosystems. Surprising and beneficial results can ensue when these organisms are reintroduced or allowed to recover.

One such group of animals potentially beneficial to freshwater fishes are beavers. Beavers are 'ecosystem engineers', their damming, canal-making and ponding activities changing the dynamics of river and catchment ecosystems, offering a natural mechanism for restoring degraded streams. Restoration of North American beavers (*Castor canadensis*) has been long-running, and has had an overall positive influence on biodiversity, creating and diversifying habitats in the US by creating ponds and wetlands, altering sediment transport processes, importing woody debris into aquatic environments and creating important habitat. Reintroductions of North American beavers also enhances stream habitat beneficially for endangered Chinook salmon (*Oncorhynchus tshawytscha*) and steelhead salmon (*Oncorhynchus mykiss*) in the Pacific Northwest, in the Upper Columbia River Basin in eastern Washington state. North American beaver introductions have consequently become a salmon conservation tool in habitat-depleted rivers in the west coast of the US. In Britain, the Eurasian beaver (*Castor fiber*), a much smaller species once common across Europe but hunted to near extinction, including its eradication from Britain 400 years ago, is also an 'ecosystem engineer' but at a much smaller scale than its American cousin. Eurasian beaver introductions have been occurring in Britain since 2011, in large enclosures in southern England as well as in Scotland. The activities and outcomes of beaver introductions in Devon have been closely monitored and have resulted in a diversity of beneficial outcomes, including significantly increasing water

**Eurasian beavers modify river ecosystems with many benefits for water quality, hydrology and biodiversity, including diversification of habitats beneficial to freshwater fishes. (Image credit: Annette Shaff/Shutterstock.com.)**

storage and quality (particularly, reducing suspended sediment, nitrogen and phosphorus concentrations), buffering of flows through increased water retention, enhancement of baseflow in rivers during dry conditions and improvement of biodiversity, including the condition of native brown trout. Review of Eurasian beaver introductions in Scotland confirms their role as 'ecosystem engineers', their activities contributing to reducing flash flooding (rather than contributing to flooding as had been claimed in the media). Whilst some fishery managers express concern that beaver dams may block passage for migratory fishes, the reality is that the Eurasian beaver does not build massive dams like its American cousin, but rather diversifies habitat to the net benefit of these species. Eurasian beaver introductions could become a conservation tool with a range of fishery, ecological, water quality and water flow buffering advantages.

Another species once numerous but now extinct in Britain is the Eurasian wolf (*Canis lupus*, or *Canis lupus lupus* as a sub-species of the grey wolf). British wolves were finally exterminated around 1680 through a combination of deforestation and active hunting rewarded by a bounty system. Various initiatives have explored the feasibility of reintroducing wolves into Scotland as a part of a wider 'rewilding' movement. But what has this got to do with freshwater fishes? Where grey wolves have been successfully reintroduced into several areas in the northern Rocky Mountains of the US, including

in Yellowstone National Park since 1995, they have been found to significantly change
the character and functioning of American wilderness ecosystems. Benefits from wolf
reintroductions in Yellowstone were almost immediate, controlling numbers, increasing
health and changing behaviour of wapiti (a deer: *Cervus canadensis*), which spent less time
in valleys and gorges where wolves easily ambush them. This promoted re-establishment
of river corridor flora, increasing biodiversity and providing food and shelter for a growing
variety of plants and animals. It also decreased riverbank erosion, stabilising and
deepening channels, with benefits flowing downstream to people. 'Trophic cascade',
where re-naturalisation of a top predator triggers profound effects throughout the entire
ecosystem, provides multiple, interconnected benefits. It remains to be seen whether we
have the vision and courage to proceed with reintroductions of species such as grey wolves,
Eurasian lynx (*Lynx lynx*) and Eurasian brown bear (*Ursus arctos arctos*) that we drove to
extinction centuries ago, of course with suitable precautions to protect lives and interests
in our now vastly changed and densely populated landscape. However, if they contribute to
the overall improved functioning of our landscapes, catchments and freshwater ecosystems,
fishes and other wildlife seem only likely to prosper as a result.

## 11.5 GOOD RELIGIOUS INTENTIONS THAT CAN TURN BAD

A scientific study I conducted with colleagues (see reference to a paper by Everard, Pinder,
Raghavan and Kataria cited in the 'Further reading' section at the end of this book) addressed
potentially serious negative consequences arising for aquatic wildlife from the inherently
well-intentioned pro-conservation and humane Buddhist practice of 'live release'. Live
release is also known by alternative names including fang sheng, prayer animal release,
mercy release, or Tsethar in the Tibetan tradition. It entails the release into the wild of
captive animals, particularly those destined for slaughter. It puts the ideal of compassion into
practical action and also, in part, compensates for the inevitable collateral killing of organisms
as humans walk, breathe and conduct their lives. Though founded on the good intention
of protection of living organisms, live release in practice also now represents a potentially
substantial pathway for introduction of non-native and potentially invasive species, which
may have perverse outcomes for the conservation and stability of ecosystems into which
they are released.

Unfortunately, major businesses have become established supplying large numbers of animals
for live release, and adherents also tend to go to markets and pet stores to buy animals for
release. Many released animals are poorly adapted for survival in the wild, including examples
of ornamental strains of birds and fish and are rapidly taken by predators or else suffer
adversely in environments to which they are not adapted. This is one perverse outcome from
an ostensibly humane activity. More worrying still has been the widespread introduction of
invasive species through this practice.

The attributes of many species suited to production in aquaculture are almost exactly consistent with those predisposing aquatic organisms to become invasive. The purchase and subsequent release of organisms for live release produced through aquaculture has consequently become a significant pathway for introduction of alien invasive species, posing conservation threats to invaded ecosystems. Serial introductions of a few widespread, non-native species globally that are readily raised in captivity exacerbates pressures driving homogenisation of freshwater fish communities. Although only one legal case – release of a non-native marine lobster species – has resulted in prosecution in Britain to date, the practice of 'live release' is a widespread traditional practice globally conducted with the best of intentions by adherents of many religions of Asian origin, including both Buddhism and Taoism. Significantly, it is part of all schools of Buddhism: Theravada, Mahayana and Vajrayana.

There are many examples of live release associated with conservation concerns, and there are sometimes legal consequences. Tsethar practices are arising as a significant concern in Bhutan, an exceptional region for freshwater fish and wider biodiversity, where the highly invasive African sharp-toothed catfish (*Clarias gariepinus*) is imported live from Bangladesh and sold for release by religiously inclined Bhutanese people. In Yunnan province, China, introduction of two species of non-native weatherfishes (*Misgurnus anguillicaudatus* and *Paramisgurnus dabryanus*) through the practice of 'prayer animal release' and their subsequent increasing populations is putting at risk the threatened native freshwater fish *Ptychobarbus chungtienensis* in Shangri-La region. A 2001 census recorded that 1.9% of the Australian population were Buddhists, also providing anecdotal evidence suggesting that purchase and release of aquarium species for live release was not uncommon, albeit entirely unquantified. Unregulated mercy releases have also resulted in invasive red-eared slider turtles (*Trachemys scripta elegans*), native to central America but widespread in pet shops across the US, now dominating and outcompeting native terrapin species in New York City's Central Park. There are many more examples in the scientific paper cited in the 'Further reading' section at the end of this book.

The point of this overview of potential adverse consequences from live release practices is not to denigrate the intentions of this inherently pro-conservation and humane practice, which are in fact welcomed and applauded. However, it is essential that practitioners are aware of these risks if their actions are not to work diametrically against the pro-conservation and humane intents of the practice. Ensuring that live release occurs safely necessitates awareness-raising and guidance informed by science to ensure that good intentions do not result in perverse, environmentally destructive outcomes. To promote this purpose, the scientific study proposed four simple principles (see Box 11.2) to ensure that the most laudable of religious intentions to safeguard the natural world do not, perversely and inadvertently, achieve exactly the reverse outcome.

## 11.6 THE UNFORTUNATE LEGACY OF 'ACCLIMATISATION SOCIETIES'

We might understand the term 'acclimatisation' today as adaptation to new climates or changing environments. However, the term had a rather different meaning during the empire-building age of European nations. Invading forces had scant interest in adapting to local environments and cultures but wondered why these new places didn't house all the same species with which they were familiar at home. These where pre-Darwinian days, when people did not understand evolution and thought instead that all species were created in some fixed state. They therefore took it upon themselves to import familiar species around the world to fill in these impoverished new lands, evidently in their eyes not so favoured by the Creator. It is this same patronising view that saw the native people of other lands as needing 'civilising', whilst exploiting their natural resources to feed the insatiable hunger of early industrialisation at home.

So, instead of adapting to and understanding these altogether different ecosystems and cultures, our colonial forebears instead set about, with not a little religious zeal, 'enriching' these ecosystems and cultures into something altogether more familiar. In other words, they 'acclimatised' both ecosystems and civilisations to an imposed European model of culture, agriculture, religion and ecology. This may seem staggeringly naïve from our more ecologically informed world view today. But export of diverse species was seen as a duty to 'fill in the gaps' relative to ecosystems smiled on more benignly by God, which of course were those of the European homelands. It was to organise these activities that 'acclimatisation societies' were formed.

The first acclimatisation society, *La Societé Zoologique d'Acclimatation*, was founded in Paris in 1854 by French naturalist Isidore Geoffroy Saint-Hilaire. Saint-Hilaire had published a treatise in 1849 'Sur la Naturalization des Animaux Utiles' ('On the Naturalization of Useful Animals'), which had urged the French government to introduce and selectively breed from foreign animals, both as food and for pest control. Later shortened to *La Societe d'Acclimatation* (the Acclimatisation Society), this group and its underlying ethos inspired the formation of similar groups around the world, most particularly amongst European colonial powers.

With the benefit of hindsight, we can clearly see how catastrophic have been the long-term consequences of many of these unwanted introductions.

British expeditions to Australia sought to remedy the strangeness of these new surroundings by introducing familiar plants and animals to make the alien environment feel more like home, to beautify their gardens, to provide sport for hunters but, above all, to make the land economically productive. The Victorian Acclimatisation Society was founded in 1861 by a private collector whose motto was 'If it lives, we want it'. It is this society that founded Melbourne Zoo, not merely as a visitor attraction but to house imported animals prior to release. Many of these introduced animals and plants – foxes, cats, blackberries but also various species of British freshwater fishes such as roach and tench – have become noxious pests ever since, threatening the integrity of Australia's fragile and unique ecosystems. Roach and tench still prosper today, particularly in the Australian state of Victoria. Native Australian fish species generally fare badly, particularly in the Murray-Darling Basin where common carp have driven the native Murray cod (*Maccullochella peelii*), a predator and Australia's largest exclusively freshwater fish, to extreme scarcity (Critically Endangered on the IUCN Red List).

# MANAGEMENT OF FRESH WATERS FOR FISH

Management plans focused on various species and groups of plants or animals are familiar in the world of nature conservation. A specific focus has strengths, particularly in relation to rare and vulnerable species with exacting requirements. However, habitat management undertaken to account of the unique requirements of individual species may obviously clash with the needs of different organisms. Natural processes create conditions to which species adapt and that consequently define their needs. The inherent dynamism of these processes is vital for maintaining ecological diversity, variability and succession. Management of processes shaping habitats and ecosystems is often more important for providing dynamic 'homes' for nature, rather than targeted, species-specific action (apart from a few of the most threatened species).

## 12.1 FISHERY MANAGEMENT FOR CONSERVATION

Management plans for fish other than those scheduled under nature conservation measures are, for reasons referred to a number of times, often seen as something other than promotion of ecosystem health. Often, they are regarded by the wider conservation community as for selfish 'fishing' purposes rather than a 'nature conservation' matter. This distinction may have been justified under an outmoded fishery management paradigm based on addressing symptoms rather than causes. For example, stocking with fish or culling of predators were often formerly default options for making good perceived shortfalls in stock size and quality, even if the reality was that external conditions, be they driven by natural or human influences, were major causes of the perceived poverty of the fish population.

However, fishery management has for a long time been going through a transition. Rather than seeking short-term fixes related to symptoms of low perceived fish stocks, there has been a progressive transition towards proactive ecosystem management to build up the resilience of fish stocks through measures such as habitat enhancement. This more strategic approach addresses the needs of mixed fish populations and the ecosystems of which they are part, and upon which they and other aquatic life depends. Rather than directing management solely at individual species of fish or other organisms, it makes sense to think systemically about the ways in which management of ecosystems can address all co-dependent organisms – mammals, birds, reptiles and amphibians, fungi, plants and other wildlife – and the many ways in which they in turn support human wellbeing. All of these facets of ecosystems are intimately interconnected and therefore ultimately thrive or decline together. Under a more integrated, cross-functional approach, significant synergies and cost savings can result from pooling expertise, budgets and political influence. Fish are important and interactive parts of aquatic ecosystems and also of the wider waterside experience, including its diverse wildlife, that is as central to enjoyment of angling as the capture of fish. There is then a progressive convergence amongst more forward-thinking people in formerly divided 'conservation' and 'fishery' interest groups, as indeed in other policy areas (such as urban planning, recreation, flood management, water quality planning and many more), about ecosystem-based solutions yielding multiple, connected benefits. Fish, like birds, enjoy a high public profile that increases support and the economic case for protecting aquatic ecosystems, whilst thriving ecosystems are naturally more sympathetic with healthy, self-regenerating fish populations and the host of societal benefits that they provide.

Of course, not everyone sees this potential convergence in quite the same way. One case in point here is seen in enclosed, commercial fisheries that may be overstocked relative to natural biomass levels, particularly with non-native fishes such as common carp and their hybrids. There is also justification for protecting some nature reserves largely for particularly vulnerable organisms, such as sites used by migrating waterfowl and wading birds that are highly susceptible to disturbance. An intensive commercial fishery is best regarded as a crop rather than as a balanced ecosystem, much like a field of maize or a flock of chickens that requires intensive management to exclude predators, treat diseases and manage other threats rather than letting natural checks and balances rebalance excessive and homogenous populations.

This chapter focuses on an ecosystem-based approach to management. This takes, as its central focus, the integrity and functioning of the whole habitat, supporting all forms of wildlife including the diverse requirements of different fish species and life stages. It also accounts for the many values that flow from them for society at large.

## 12.2 FISH STOCK MANAGEMENT

Unconstrained commercial, recreational and subsistence fishing has the potential, if exceeding natural production rates, to threaten freshwater fish populations. A wide range of measures has consequently been implemented to maintain the viability and health of mixed freshwater fish populations. These include various forms of restrictions on fish exploitation as well as a range of other proactive measures.

### 12.2.1 Method-based restrictions

Some fishing techniques are potentially destructive, resulting in irreversible damage to aquatic habitats and ecosystems if used inappropriately. Many of these are illegal or else are regulated. Part I of the Salmon and Freshwater Fisheries Act 1975 is headed 'Prohibition of Certain Modes of Taking or Destroying Fish etc.', setting a range of restrictions. These include prohibition of a range of destructive methods including explosives, poisons and firearms, taking or selling immature fish or those about to spawn, restrictions on types of nets and making it an offence to discharge polluting effluent. Set lines of baited hooks are also generally illegal, as they often result in fatalities both to fishes and to other wildlife, as well as representing potential human hazards.

Some fishing methods require permitting to ensure that their use is controlled. For salmon and trout fishing, these include various forms of nets (compass nets, haaf nets, draft hand and trammel/whammel nets, wade nets, coracle nets, T&J nets (fixed engines placed at right angles to the beach), drift nets, gill nets and seine nets). For eels, fyke nets, dip nets, pots, baskets and criggs require a licence.

**Fyke nets are one of many forms of net used to harvest or sample fish, and require permitting to control fish stocks and wider impacts on aquatic ecosystems.**

### 12.2.2 Geographical restrictions

There is also a range of geographical restrictions. For example, Part II of the Salmon and Freshwater Fisheries Act 1975 is headed 'Obstructions to Passage of Fish'. Fixed engines, unauthorised fishing weirs, taking migratory fish and placing restrictions on taking salmon or trout above or below certain obstructions are illegal (including, for example, fishing near fish passes). There is also a requirement to ameliorate the effects of obstructions to fish passage. Netting of migratory fishes is permitted in estuaries, but only within prescriptive licencing conditions.

Although angling on still waters has, since the late 1990s, been permitted year-round across Britain, landowners and nature conservation agencies can restrict angling on some waters, or zones of large still waters, where there is potential to disturb sensitive wildlife. On many recreational fisheries, 'no fishing' zones are established electively where fish can find refuge or which are important breeding or nursery areas.

### 12.2.3 Temporal restrictions

One of the better-established mechanisms put in place to protect fish populations during vulnerable life stages is the imposition of statutory or elective closed seasons for both recreational and commercial fishing. Different closed seasons now apply to differing types of fishes – coarse, salmon and trout – typically during their most vulnerable breeding periods. Social protocols may have served this purpose previously, but the origin of some of the law on this matter is of interest in terms of its social, as well as environmental, context.

One such restriction is the instigation of the closed season for angling for coarse fish species in England and Wales. This has its roots in the latter half of the nineteenth century, when growing concerns about potential depletion of freshwater fish populations led Anthony John Mundella to petition for and drive through legislation relating to their protection. Though little known today, Mundella rose from humble beginnings as the son of Italian refugees, rising through a family

Weirs can present impassable barriers to fish and so require passes to enable fish movement up and down river systems. (Image © Dr Mark Everard.)

role as a hosiery manufacturer to become a prominent Liberal Party politician and reformer. Mundella was first asked to stand for parliament to defend the interests of labour in the wake of the Sheffield Outrages, a civil uprising by working classes upon whom were inflicted the worst demeaning and health-threatening conditions of the Industrial Revolution. Mundella sat in the House of Commons from 1868 until his death in 1897, including two stints as President of the Board of Trade. During this time, he made numerous prominent contributions to compulsory education, animal health, worker' rights and removal of restrictions on trade unions. Mundella was also concerned with the wellbeing of freshwater fish stocks. Prompted by a consortium of seven thousand or so anglers in his constituency of Sheffield, Mundella drove through proposals '...which will permit of freshwater fish completing the procreation of their species in peace and *quiet'*. Vociferous debates ensued about the time period this should cover, noting both the varying spawning times of different coarse fish species ('game' species, such as salmon and trout, spawn in the winter and so are covered by different closed seasons) and also considerable variation due to inter-annual and geographic factors. But, to cut the long and sometimes acrimonious debate short, the consensual period spanned the middle of March to the middle of June as a compromise, splitting the difference between a Sheffield proposal of March to May and a London counter-proposal of April to June. This formed the basis for the coarse fishing closed season from 15 March to 15 June (inclusive) enshrined in the Freshwater Fisheries Act of 1878 (the 'Mundella Act') and rolled forward on rivers for coarse fish angling to this day across England and Wales. Every year, there are representations – mainly economic or from a sense of entitlement divorced from concern for fish ecology – for the river coarse fishing closed season to be removed. (The battle was lost in the late 1990s on retaining the closed season on still waters, subject to the discretion of landowners and nature conservation agencies, again primarily driven by economic reasons.) However, whilst it is clear that a fixed ninety-three-day span may not cover the spawning period of all species every year in all corners of the country, the basic tenet that coarse fishes deserve a period of no disturbance when they are at their most vulnerable is, to my mind, entirely warranted. No such closed season for coarse fishing exists in Scotland or Northern Ireland, where controls on angling are at the discretion of landowners.

Conservation of salmon and trout stocks during their breeding period is also the underpinning rationale for angling closed seasons, though these are variable across the country, reflecting local breeding patterns. For example, down in the relatively warmer south of England, the Atlantic salmon angling season on the River Itchen (Hampshire) opens on 17 January and runs through to 2 October, with the sea trout angling season running from 1 March to 31 October. On the flashier River Axe in Devon, the fishing season for salmon does not start until 15 May, running through to 31 October, with sea trout angling permitted from 15 April to 31 October. In Scotland, the fishing

**Time-related restrictions on fishing activities are common means to protect freshwater fishes during vulnerable periods, such as the mid-March to mid-June closed season for recreational angling on rivers in England and Wales.**

season for wild trout generally starts on 15 March and continues to 30 September, with some variations, though in Aberdeenshire and the Highlands, the season generally starts on 1 April to give the trout a couple more weeks to return to good condition after the colder winter further north.

Similar, locally adapted closed seasons are applied as conditions of licences for commercial netting and other forms of fishing.

### 12.2.4 Population-based restrictions

An important control put in place to protect unique genetic strains and secure escapement of sufficient spawning fish concerns mixed stock fisheries of Atlantic salmon in high seas fisheries. Serious damage to Atlantic salmon stocks resulted from discovery by commercial netting interests in the 1980s and 1990s of the feeding grounds of adult salmon off the coasts of the Faroe Islands and West Greenland. Atlantic salmon originating from most of the Northern Hemisphere's countries congregate in these important feeding grounds, constituting a truly mixed stock. Exploitation of these high seas fisheries therefore resulted in damage that, though hard to quantify, can be assumed to be substantial, as it did not discriminate stock from any particular country and particularly from any individual river. Exploitation consequently bore no relation to whether natal river systems held healthy or struggling Atlantic salmon populations. Efficient management of salmon stocks was therefore seriously threatened.

Action to control the indiscriminate, potentially irreversible damage caused to the whole Northern Hemisphere stock of Atlantic salmon was driven by the North Atlantic Salmon Conservation Organisation (NASCO), formed in 1989 as a result of a resolution taken at an Atlantic Salmon Trust conference in Edinburgh. NASCO's original brief was to develop catch quotas for the Greenland and Faroese fisheries, providing national governments with greater opportunities to manage their stocks including the far from insignificant socioeconomic benefits associated with

**River-by-river conservation of Atlantic salmon is an important measure to protect specific local populations.**

rod-and-line angling and net fisheries. However, under the NASCO umbrella, a range of non-governmental organisations (NGOs) led by Orri Vigfusson, an Icelander who had become wealthy through business interests and then diverted his considerable energies, until his death in 2017, to conservation of Atlantic salmon, developed the North Atlantic Salmon Fund (NASF) to provide compensation for Greenland and Faroese fishermen. Both political agreements and commercial compensation packages combined, often in an uneasy way, to promote conservation of Atlantic salmon stocks on their marine feeding grounds. Under these agreements, the Faroese have retained their right to fish for Atlantic salmon on the high seas, but have not exercised that right since 2000, whilst Greenland has agreed a subsistence tonnage for home consumption. NASCO monitoring teams take samples from Atlantic salmon caught from the Greenland fishery, with advances in detection of genetic markers of fish from individual river systems providing invaluable information as to where different national stocks feed.

Whilst high seas fishing was relatively recent, it has been subject to commercial controls. Consequently, the focus has returned to the coastal fisheries of Northern Hemisphere countries to take proportional responsibility for their native and local stocks.

Coastal netting of Atlantic salmon has taken place for centuries. Scottish fishermen opened a drift net fishery in the early 1960s. However, following a recommendation in the 1963 Hunter Report, the fishery was promptly closed without compensation for netsmen. The Republic of Ireland took the bold and strategic step of closing its drift net fishery. In England, the large drift net fishery off the northeast coast, which once killed more than 25,000 fish a year, was significantly reduced in 2003 in a buyout partnership between the government and private fishery interests. Netting of Atlantic salmon and sea trout still continues at lower intensity on the northeast English coast, a concerning trend at a time when both Atlantic salmon and sea trout stocks are under pressure throughout the UK. The Republic of Ireland and the northeast England drift net fisheries were relatively easy to close and reduce, respectively, albeit over extended timescales and entailing much negotiation. This was achieved principally because they were both licensed and controlled by governments. In Scotland, however, rights to fish are heritable, and therefore the government has less control over management options, though effective legislation is available if the political will can be found to apply it. One can only hope that research currently being undertaken into sea trout life cycles will inform better management policies in the future.

In Scotland, 11,738 Atlantic salmon and grilse and 8,663 sea trout were reported as caught in 2010 in estuarine and in-river net and coble fisheries (cobles are sweep nets shot in a wide circle from beach or bank using boats). However, due mainly to local buyout initiatives and, hence, a greatly reduced netting effort in estuaries, this is a mere fraction of historical levels, albeit still significant. Also, these netting stations come under the management policies of individual District Salmon Fishery Boards and so can be regulated, along with rod and line fisheries, according to local stock status.

In England and Wales, again, buyout schemes have greatly reduced netting effort, and only a small proportion of the English and Welsh net catch now occurs in estuaries. We have probably reached a satisfactory level of exploitation with estuary netting, because it would be sad to lose completely the heritage value of salmon netting sustaining our forebears for centuries. This measure, together with a fish carcass–tagging scheme which means that all wild fish sold are traceable, is believed to substantially reduce the legal and illegal taking of sea trout, better to align it with sustainable levels of exploitation.

The final word on all commercial netting, whether on the seas, in coastal waters or in estuaries, is that all fish have to die to contribute to local socioeconomics, and this inevitably has management implications. Rod and line fisheries can still contribute strongly to local economies while exercising conservation restraint by adopting a catch-and-release ethos which, with the high survival rate achieved in most fisheries, adds greatly to spawning escapement and the future wellbeing of the stock.

### 12.2.5 Catch-and-release

Marking the transition of angling from a means of gathering food to a recreational activity informed by stock conservation concerns, catch-and-release fishing has become widespread. Coarse fisheries have been generally run on a catch-and-release basis for many years, at least since the Second World War, during which time taking coarse fish for the table provided families with a valuable protein supplement. However, the catch-and-release ethos was slower to take root in game angling. Nevertheless, concerns about stocks of salmon, in particular, have given rise to a significant shift in opinion in attitudes and practice throughout the first decades of this century.

The take-up of catch-and-release Atlantic salmon fishing is now enforced regionally by byelaws. For example, the Spring Byelaws, brought in by the Environment Agency in 1998, prohibit the killing of salmon before 16 June to protect early season stock (generally larger, multi-sea-winter fish that stay at sea for two, three or more years before returning to breed). Surveys reveal that, despite widespread objections to compulsory catch-and-release byelaws at their inception, anglers' attitudes to conservation had changed so markedly by 2008 that there was a high degree of support to keep the initiative alive for a further ten years. Even those rivers with an arguable surplus of early running salmon, such as the Tyne, voted to keep to the byelaws, mainly because of concerns that the river would be inundated by anglers trying to catch a springer 'for the pot' should this river alone be granted dispensation. Definitive figures are elusive. However, the transition is marked by a survey reported in the Environment Agency's *National Trout and Grayling Fisheries Strategy* that half of the sea trout caught by anglers were released in 2000 compared with 29% in 1993. A 2001 survey of licence-holders indicated that, of those who fished mostly for them, 78% always or usually practised catch-and-release for wild brown trout.

Catch-and-release angling was slow to be taken up by game anglers but is now widely embraced and appreciated, enabling released fish to contribute to breeding and sustainable stocks. (Image © Dr Mark Everard.)

Good fish care is, of course, essential. However, survival rates after capture and release by anglers are very high, ensuring that released fish can continue to pass on their genes. The release of fish also contributes to sustainable angling tourism industries, which may be locally very significant.

This trend towards catch-and-release has been bucked by some sectors of society for whom the capture of fish to eat is a cultural practice. Eastern European people, in particular, have grown up with the expectation of keeping the fish they catch, and also fishing without licences or otherwise illegally, largely through ignorance of British laws and customs. The Angling Trust, the major governing body for angling in Britain, responded to increases in illegal fishing observed in a number of areas through a 'Building Bridges' project, part funded by the Environment Agency, to improve understanding in Eastern European communities of the byelaws and angling culture of this country. Posters in multiple Eastern European languages as well as anti-fish theft signs are also made available for download from the Angling Trust website, and can be seen displayed on many waters controlled by angling clubs.

### 12.2.6 Protection of genetic diversity

As outlined in the chapter 'Conservation', the genetic integrity of locally adapted strains of fishes has been under long-term threat from stocking and other forms of fish introduction. Measures are being progressively put in place to protect native stocks. A government advisory body, the Moran Committee, representing angling interest in England and Wales, noted that '*Wild brown trout populations… are… at risk, not only through displacement by stocked fish, but also by interbreeding with stocked fish…*' A range of protective measures has been subsequently recommended and increasingly implemented by protocol as well as through byelaws. One such strategy is published in the form of the Environment Agency's 2003 National Trout and Grayling Fisheries Strategy, which states that

> *In general, we will continue to consent stocking into 'native trout' waters if: consistent with practice over the last five years (the objective is to avoid increasing stocking levels of fertile, farm strain trout); or stock fish are triploid females; or stock fish are derived from local, naturally produced, broodstock under a suitable rearing regime.*

In other words, stocking with trout that may dilute the native strain will implicitly not be authorised. In an update newsletter published in July 2010, the Environment Agency strengthened this guidance to state that

> *One element of the Strategy is to discontinue the stocking of fertile farm strain (diploid) brown trout into rivers and other unenclosed waters by 2015. Instead, stocking should be with non-fertile (female triploid) farm reared brown trout or the progeny of local brood-stock reared under a suitable regime. Up until 2015 we are encouraging a voluntary switch to female triploid brown trout stocking with the aim of seeing a 30% fewer fertile farm strain brown stocked by 2010 and a 50% fewer by 2013.*

Brown trout and rainbow trout are stocked all over the world, as are many other freshwater fish species, generally from a limited or even a hybridised gene pool, and fish species are also widely transferred between drainage basins or countries. The genetic mixing and potential for negative consequence of these actions, all part of a wider global genetic homogenisation, is at present poorly appreciated and reflected in controlling legislation. One can only hope that the associated risks become better understood and, consequently, controlled.

### 12.2.7 Stocking versus natural recruitment

The stocking of fish in British waters has been an established practice over many centuries, with our fish fauna being perhaps the most disturbed of all taxonomic groups. This may have involved introductions of species initially for aquaculture and food purposes, which has seen the common carp introduced into British waters and indeed many waters across the world, often with severe environmental consequences. This practice continues today to serve lucrative ornamental fish and angling industries. Other species have been moved for similar purposes beyond their native ranges. For example, barbel and grayling are now found in many British rivers despite being naturally distributed towards the south and east of England.

The grayling has been introduced widely across Britain beyond its native range, largely by angling interests.

Stocking of fish can have worrying conservation implications. For example, introduced species can become established and compete with locally adapted species (as is the case with barbel). They can also modify the ecosystem through their feeding habits (frequently observed worldwide with common carp introductions) or prevent the breeding of native species (a major concern with introduced topmouth gudgeon and sunbleak). Introduced fish can also carry novel diseases (a major concern in transfers of farmed salmon) or unwanted 'stowaways' such as other invasive fish species (such as ruffe), problem plants (such as New Zealand pigmyweed, *Crassula helmsii*) or other animals (including the 'killer shrimp', *Dikerogammarus villosus*).

However, stocking of trout, significantly often including the non-native rainbow trout, can be a positive conservation measure when angling pressure would otherwise deplete natural stocks. Stocking is of course no panacea, potentially giving rise to many of the introduction-related problems discussed above but also risking the dilution of localised genetic strains of trout and salmon. It is for this reason that policy changes are encouraging a progressive transition either to stocking with native provenance brown trout or else with increasing proportions of sterile (generally triploid) brown or rainbow trout to avoid genetic impacts upon native stocks. Optimally, ecosystem-based management can improve spawning and recruitment success for native fishes, reducing or entirely averting a need for stocking.

### 12.2.8 Controls on poaching

Despite increasing controls on some forms of fishing, both commercial and recreational, poaching (the illegal capture of fishes) – particularly of Atlantic salmon and sea trout – remains a problem in Britain, as indeed elsewhere that salmon species run. The clandestine nature of this activity makes it impossible accurately to quantify the illegal harvest of British game fishes. There is evidence of decreasing poaching activities in recent years, both due to depleted runs of wild fish and the cheap price of farmed fish. However, illegal fishing continues and can cause specific problems locally which, with the diminishing resources available to statutory authorities for enforcement, is often difficult to police. This has become somewhat easier in England and Wales since the beginning of 2009 with the passing of a national byelaw prohibiting the sale of rod-caught fish and, most importantly, introducing a carcass-tagging system for commercially caught Atlantic salmon and sea trout. Any fish on the trader's slab or in the restaurant's kitchen which does not carry a unique identification tag is considered illegal. Enforcement can now be more efficiently focused at the consumer end, rather than being spent on long and risky surveillance missions in the faint hope of catching poachers 'red handed' on the river bank.

In Ireland, both north and south of the border, tagging has been successfully established for some time, and the sale of rod-caught fish was banned in Scotland under the Aquaculture and Fisheries Act 2007. However, the Scottish Government did not impose a carcass-tagging scheme, and so, one of the major recommendations from a fisheries working group operating throughout 2009 was for this to be implemented as soon as possible, closing a potential ready market for fish caught illegally anywhere in the British Isles.

However, as has been mentioned elsewhere in this book, other environmental and man-made factors come into play with poaching. For example, local fisheries interests have fought for several years to have the fish pass on the Tees Barrage made more efficient in allowing Atlantic salmon and sea trout to run through it. The main consequence of a hard-to-pass barrier is that fish stack up below it in unnaturally large shoals and become easy prey for both mammals and humans. So, the thriving population of Tees estuary seals has had a glut of fish in recent years, and a steady trickle of Environment Agency prosecutions for human poachers suggests that many fish have ended their days being consumed illicitly by people.

Likewise, if a river is closed to angling for any reason, such as occurred in the Irish Republic in 2007 as a political sop to the netsmen for the closure of their drift net fishery, the 'eyes and ears' of the anglers are lost as an informal monitoring and enforcement tool. Poachers know that, and so the very act that was supposed to conserve fish stocks ends up perversely having the opposite effect. If anglers are on the riverbank, they act as deterrents to poachers, and they attract continued investment into the river by fishery owners, while catch-and-release optimises spawning escapement. This represents a 'win-win' situation. Hence, organisations such as the NGO Salmon & Trout Conservation UK (S&TCUK) have lobbied for river fisheries to be kept open for angling if at all possible, even if this means imposing local compulsory catch-and-release restrictions.

### 12.2.9 Sampling freshwater fishes

All of the above of course assumes that you know what and how many fishes are in the water that you are managing. Without getting into distracting technical depth – there are plenty of other sources if you want greater detail on methods – it is instructive at this point to consider a few of the sampling methods used to assess fish presence and stock density.

The most passive of method is simple observation, and we cover this in a whole chapter towards the end of the book ('Fish twitching', the fishy equivalent of birdwatching or 'bird twitching'). Simple and patient observations can reveal a lot about the more conspicuous species in a water. So, too, can talking to anglers, if they are active on the river or lake. Of course, neither method is ideal for detecting all species that may be present and particularly those (such as tench, European eels and lamprey species) that tend to be more elusive and/ or nocturnal, as well as smaller species (such as stone loach and bullheads) that also tend to remain hidden and do not attract angling interest. But if a 'catch per unit effort' approach is taken, using good angling records to assess rate of capture versus angling effort, some idea of species presence and stock densities can be achieved. (Angling returns are used by the Environment Agency to monitor national Atlantic salmon and sea trout population status in England and Wales.) Other more sophisticated, passive methods include automatic fish counters (used, for example, in the monitoring of migratory salmonid fishes), using sensor technologies to log passing fish in controlled channels such as those bypassing weirs.

Progressively more active sampling methods including various forms of netting and, also, electrofishing. Seine nets, encircling a section of river or still water, can sample fishes in the water body within that area, though, again, smaller species tend to be substantially under-represented. Gill nets tend to be a destructive technique, operating, as their name suggests, by trapping fishes by their gills and generally damaging and killing them. Consequently, gill netting is more suited for commercial harvesting than for routine monitoring, and even then, it tends to be non-selective. Electrofishing entails sending an electrical current through the water between

Electrofishing is a non-destructive means for sampling fish populations in fresh water.

a submerged cathode and an anode. Fish are attracted to the anode, and the current stuns them briefly without lasting damage if the current is carefully calibrated to water conditions. As fish are stunned, they are removed from the water using a non-conductive net and handle, generally held in a bucket of some sort whilst any relevant measurements are taken (such as species, weight, length and/or health) and then released. Combining these methods, for example, stop-netting a reach of river and then electrofishing it more than once, can give a detailed assessment of species presence and population density.

Other methods for sampling freshwater fishes include sonar equipment. There are diverse types of sonar systems for this purpose, also known as fishfinders or sounders, that detect fish and other underwater objects using reflected pulses of sound energy. Some degree of experience in necessary to distinguish large shoals from big individual fish, underwater vegetation or other obstructions.

Many of these techniques are for specialists rather than amateurs, though 'fish twitching' is an art and enjoyment open to all, angling is a sport enjoyed by many, and many relatively affordable fishfinders are readily available on the market.

**Sonar-based fishfinders are increasingly cheap and readily available. (Image © Dr Mark Everard.)**

## 12.3 PREDATOR MANAGEMENT

Predation is a natural process, common to all ecosystems, by which energy and chemical building blocks are transferred between organisms. It is also one of the mechanisms by which populations of prey species are regulated.

Predator control has been a prominent feature of historic fishery management, reflective of an anachronistic, less ecologically aware and attuned age when piscivorous fishes such as pike, as well as other groups of organisms, were regarded as detrimental to stocks of other fish species. Fortunately, pike are now regarded as not only a valued angling target but also better respected in terms of their natural roles in maintaining the balance and health of mixed fish stocks. Where pike culling still persists, as, for example, in some commercial stocked trout fisheries, the outcomes, perversely, can include a net increase in pike populations, rates of predation and injury to other fishes. This is due to the highly cannibalistic nature of pike, larger pike being a primary agent of control of smaller pike. If the larger, slower-growing top predators are eliminated, dense populations of small, hungry 'jack pike' can form. These younger fish are fast growing and with a commensurate rate of predation and also with less hunting experience, which leads to them tending to injure potential prey fish that are too large or are attacked too ambitiously. Natural balances, crafted over millions of years, contain an inherent wisdom that we override at our peril.

Whilst the presence and piscivorous activities of common (or Eurasian) kingfishers and grey herons (*Ardea cinerea*) are taken as 'natural' by most people with an interest in Britain's freshwater fishes, and the little egret (*Egretta garzetta*) is equally accepted though a relatively recent natural coloniser, the same cannot be said of the great cormorant. Predation by dense cormorant populations can overwhelm the regeneration rate of fish stocks, particularly in

**Pike are highly cannibalistic, larger pike providing effective control of pike populations.**

smaller waters, though this is largely a consequence of our own making. Up to the 1970s, Britain's cormorants were overwhelmingly coastal birds, with inland sightings something of a rarity. However, coincident with intense coastal commercial fishing, severely depleting stocks of marine fish, was the creation of substantial areas of 'inland seas' in the form of flooded pits from which gravel had been extracted to feed our seemingly insatiable hunger for construction materials. The birds have understandably switched habits to exploit this new, fish-rich niche. The impacts of this are not restricted to large inland standing waters, as birds fan out over wider landscapes to feed. Predation pressure on rivers can be extreme when teams of cormorants descend upon them to feed when large still-water bodies freeze over, denying them access. Creative and consensually derived solutions are required to offset the impacts of increased cormorant predation, given current high inland populations of these birds. Habitat enhancement can be important in offering refuge to different fish species and life stages. Bird scaring and partial culls may also have a place, perhaps unpalatable to the 'birding' community but seen as a control measure not simply of narrow angling interests but of the conservation of a diverse and balanced fish fauna, perturbation of which has profound implications for diverse linked species and natural processes.

Recovery of Britain's population of Eurasian otter (*Lutra lutra*) since the nadir of the 1980s is another factor increasing predation pressures. Eradication of this charismatic predator within our shores seemed a distinct possibility, following long-term precipitous decline in numbers owing to the introduction and widespread use of various persistent pesticides in the post-Second World War period. Their recovery is a major success story due not, as some misinformed views suggest, to captive breeding and stocking. Recovery and recolonisation of the otter population, now with signs or territories in every county across Britain, was not only a predictable, but also a predicted, outcome of progressive phase-out of the more problematic,

**Grey herons are natural predators of fish, amphibians and small mammals by the waterside, though an abundance of refuge habitat helps fishes evade many types of predators.**

bioaccumulative agrochemicals. Biopsies of otter carcasses during the early recovery period of the 1990s found the predominant cause of death of this mobile mammal was road fatalities. Today, the major cause of mortality of this highly territorial animal is through injuries inflicted by other otters. Like kingfishers and robins (*Erithacus rubecula*), otters do not form dense populations, but drive off any interloper into their territory with lethal force. Also, though rapid and agile swimmers, otters do not, by preference, waste energy chasing active fishes. Rather, their stout, sensitive whiskers equip them well for probing between and under submerged rocks and in silt for more sessile prey. European eels are a favoured quarry, albeit a depleted food source today as eel populations have crashed. Various surveys of the contents of spraints (otter droppings deposited as territorial markers) reveal that small fish species, particularly including bullheads, form the overwhelming bulk of an otter's diet. Except, that is, where American signal crayfish are abundant, as otter spraints in these areas often appear bright red from pigments produced from an almost complete diet of these troublesome, invasive crustaceans. These biological realities of density-limited territorial populations and feeding preferences should allay the variously expressed but misplaced fears that the return of otters rings a death knell for Britain's freshwater fishes, or that otter populations will continue to become denser. Historical records demonstrate quite the reverse. Fish stocks were abundant in our rivers and lakes in the pre-Second World War and earlier eras alluded to by many anglers as 'Golden Eras' of rich fish stocks, all well preceding the decline of this charismatic and once ubiquitous top predator.

Various other elements of the British fauna have historically attracted the ire of overzealous fishery managers in years gone by. The bullhead, for example, was once cast as a villain in salmonid rivers as a predator of fish eggs and fry. Today, however, most, if not all, river

**The Eurasian otter naturally feeds by turning stones and grubbing on the bed and margins of water bodies, favouring prey small fishes found there but more recently also exploiting booming signal crayfish populations.**

managers have a far more proportionate perspective on the relative contributions of natural predation as compared to the major impacts of human-induced pressures.

## 12.4 SPECIES INTRODUCTIONS

In considering the many threats to Britain's freshwater fishes in the preceding chapter, 'Conservation', the dangers of species introductions beyond their natural range were made abundantly clear. Fish species and other organisms introduced, be that deliberately or accidentally, beyond the natural ranges in which they have co-evolved can run amok when liberated from their natural checks and balances, potentially changing the characteristics, functioning and quality of the freshwater and other environments into which they have been introduced.

### 12.4.1 Deliberate fish introductions

Fish have been widely introduced throughout a long history and continue to be transferred between waters. This occurs sometimes legally but also illegally, as well as accidentally.

Some stock transfer operations directly support conservation goals. For example, the European whitefish and the vendace require good quality, well-oxygenated water in deep lakes that are free from excessive egg predation. As we have seen, they face a range of threats in contemporary landscapes, including various forms of pollution, sedimentation of spawning grounds and the introduction of coarse fish species such as roach and ruffe. Ruffe have been implicated as a major factor in the extinction of vendace from Bassenthwaite Lake (Cumbria) and pose similar threats to the Welsh population of European whitefish (the gwyniad) in Lake Bala. As a conservation measure, gwyniad eggs have been transferred to Llyn Arenig Fawr, a nearby cold lake. The Scottish population of the European whitefish (the pollan) has also been

The ruffe is implicated in elimination of vendace from Bassenthwaite in the English Lake District and also threats to European whitefish (gwyniad) in Lake Bala.

successfully introduced as a conservation measure into two more sites: Loch Sloy and the Carron Valley Reservoir.

Stocked fertile brown trout have also caused problems by the genetic homogenisation and loss of locally adapted strains of the species. The widespread release and naturalisation of goldfish threatens the viability of pure strains of the closely related crucian carp, which will also interbreed with the introduced common carp. This challenge applies as much to species native to Britain that are transferred beyond their native range, as, for example, the barbel formerly native only to rivers from the Humber catchment in the north to the Thames in the south through the Ice Age land bridge with continental European rivers. Introduction of aggressively competitive species displacing native species and strains is a pervasive global problem, contributing significantly to genetic homogenisation. Introduced species can also bring with them unwanted novel parasites and diseases, best controlled by prevention rather than a curative approach that may be impossible once established.

### 12.4.2 Accidental introductions of fish and other organisms

As seen with the ruffe, hitchhiking unsuspected with larger stocked fish into the Kennet and Avon Canal and as of 2017 now spreading into the lower Bristol Avon river system, some fish can stow away unsuspected. The same may be true of alien topmouth gudgeon and sunbleak introductions and their subsequent proliferation into nuisance proportions in British waters, either stowing away with other stockings or else resulting from well-intentioned or accidental aquarium releases. In these cases, damage can and has ensued.

Also addressed in the Chapter 11, inadvertent introductions of invasive plant and animal species, from New Zealand pigmyweed to killer shrimps, the spread of signal crayfish by accident as well as intent, zebra mussels, 'rock snot' algae and a wide range of other organisms have the potential to cause damage. Best then to focus on biosecurity in all aquatic activities, be that on the boots and protective clothing of researchers and regulators working in freshwater systems, boating and angling interests, and other potential vectors of the spread of problem species. This is simply achieved by the precaution of washing and thoroughly drying, or ideally disinfecting, boots, pond nets and fishing nets or other equipment on or in which small species or their propagules or diseases may be carried. The GB Non-Native Species Secretariat (NNSS), a cross-sectoral advisory group, published a GB Invasive Non-native Species Strategy in 2008, since regularly updated. The NNSS is active in meeting the challenge posed by invasive non-native species in Great Britain. For aquatic organisms, it publishes a check-clean-dry protocol for boaters, anglers and other water users. This protocol replicates similar advice in the US and elsewhere, to take simple precautionary measures to prevent inadvertent transfer of invasive, potentially problematic plant and animal species between freshwater bodies.

Zebra mussels have become a globally pervasive freshwater invasive species, blanketing underwater surfaces and blocking water pipes. (Image credit: Jeff Caughey/Shutterstock.com.)

### 12.4.3 Other fish invasions

Concluding this thinking on fish introductions, it is interesting to consider the cases of species that are a combination of both human-vectored as well as 'accidental' introductions. As addressed in preceding chapters, the pink salmon is native to rivers in North America draining to the Pacific Ocean. However, following introductions of the species by Russia into the Barents Sea, pink salmon have been making progress around northern Eurasia, appearing and establishing in Norway and Iceland. A few largely unsubstantiated reports of pink salmon have come from Scotland and northern English rivers over recent years. As of 2017, pink salmon have been filmed spawning in the River Ness in Scotland, and a significant number of verified fish have since been caught by anglers from Scottish and northern English rivers, including the Tyne, as well as in Irish rivers as far south as Galway. The progressive invasion of British rivers by the highly fecund pink salmon seems inevitable and the consequences for species competition and other outcomes is yet to unfold.

Another species that we can expect to make a reappearance could be the decidedly uncommon and enigmatic common sturgeon. This is due to systematic propagation of the

The pink salmon is a recent invasive species, appearing and locally spawning in British rivers. (Image credit: COULANGES/Shutterstock.com.)

species in France since 2007, as a desperate measure to conserve a species for which natural reproduction seemed to have ceased. Sturgeon are very long-lived fish, and where they go during their long adult marine phase remains a mystery. However, although it is unclear whether common sturgeon have been occasional visitors to estuaries and lower rivers in Britain or if they actually bred, we should not be surprised if some individuals from this French-bred stock eventually appear in some of our larger estuaries.

## 12.5 HABITAT ENHANCEMENT

Whilst we may think of many of them as amongst the most 'natural' features of our much-manipulated landscape, in reality, the habitat of Britain's fresh waters is substantially impoverished through direct conversion and the multiple pressures of catchment land use. This, in turn, poses a threat to aquatic wildlife, including creating 'bottlenecks' in the life cycles of freshwater fishes. As discussed in detail in earlier chapters of this book, different freshwater fish species, and critically also their various life stages, have specific and interdependent feeding, refuge, breeding and nursey habitat requirements. A thriving aquatic ecosystem can support multiple species of fish amongst an array of other water plants and animals, whereas one that is habitat-poor is likely to be dominated by a smaller range of species or, potentially, by large gaps in 'year classes' of fish and other organisms, reflecting poor recruitment in anything other than ideal climatic conditions in former years. A squeeze on estuarine habitat through various forms of port, resort, urban, industrial and agricultural encroachment and development can also limit habitat available for fishes dependent on transitional waters for completion of their life cycles, including breeding, feeding and an adequate adjustment period when moving between fresh and saline waters.

Where rivers, lakes and other fresh waters are impoverished, habitat improvement measures can go a significant way towards relieving these bottlenecks. It can also make an often

**River habitat diversity is good for fish populations and other wildlife.**

substantial contribution to alleviating broader problems. For example, restored marginal and riparian habitat can halt or slow the flow of contaminated runoff entering water bodies, support microorganisms that process pollutants and thereby purify the water, and slow flows, resulting in settlement of suspended sediment. Habitat diversification can also compensate to a degree for low flows and water levels, especially where vegetation and some structures hold back the water level and constrain river channel width, creating flow diversity and bed scour, and ensure that sufficient marginal features remain in still waters. Habitat diversity in all cases offers refuge from increasing levels of predation and from spate flows.

Tackling major threats from pollution, water abstraction and damage from large developments are things that we can try to influence remotely. However, issues of biosecurity and the protection or enhancement of habitat are things that we can engage in actively for the improvement of conditions for fishes and other freshwater life. A few ecosystem-focused guides are available to inform these activities, including *The Wild Trout Survival Guide* and *River Habitats for Coarse Fish: How Fish Use Rivers and How We Can Help Them* (listed under 'Further reading' towards the back of this book). An overview of some key areas of proactive habitat conservation work is provided here.

### 12.5.1 Valuing natural assets

Often, a working party on a pond or river, assembled by a community association, angling club or other group, will focus on making a water body visible and accessible. The chain saws and secateurs are brandished with all good intent, but often without thinking about what the fishes, birds, insects and other biota require in addition to the human desire for grand vistas and tidiness. Many is the freshwater ecosystem denuded by good intentions! So the first advice is to take account of the habitat available for all fish species present or desired, including their feeding, breeding and refuge needs. Is that 'untidy' sallow bush dipping its boughs into the water in reality a major asset for all these needs? What would be the consequences of its removal? It is best to walk the banks of the water body as a small team, bouncing around ideas about how the habitat supports the three primary needs of fishes – feeding, breeding and refuge – across all life stages. Then, to map habitat assets that may have been formerly overlooked as they were not considered in these terms, marking them down as features to protect. This generally results in best outcomes for all freshwater life, not just fishes.

A nicely vegetated river margin, and bankside bushes marching or dropping into the river's edge, are actually a precious resource rather than a hindrance, as they provide for many of the habitat needs of a range of fish species. So too are mid-channel reed beds creating slack water at their downstream ends and accelerating flows and bed scour to either side. Beds of submerged plant species such as the underwater 'cabbage' leaves of yellow water lilies (*Nuphar lutea*) also provide friction to slow flows, creating feeding niches and slack water for fry and small species. For lilies in particular, the rhizomes are perennial, providing valuable feeding, nursery and refuge habitat year-round, even as the leaves die back in winter.

Waterside trees and their root systems confer many benefits on rivers and other freshwater ecosystems, diversifying flow regimes and habitat, stabilising banks and also offering shade, refuge, feeding and breeding habitat. (Image © Dr Mark Everard.)

Waterside trees such as various species of willow (*Salix spp.*) and alder (*Alnus glutinosa*) can be important not merely for habitat provision by dipping boughs and through dropping insects and other food into the water, but as they may produce dense underwater balls of roots that not only strengthen the bank but provide perennial refuge, feeding, nursery and potential spawning habitat. These riparian trees are also a valuable source of shade, contributing to an often significant reduction in river temperatures, as well as providing cover from aerial predators.

So the first advice concerning habitat management for the protection or improvement of fresh waters for fish populations and other wildlife is to be aware of what is already there that is of functional value and to mark it for protection rather than 'tidying it up' or managing the riparian zone purely to improve human access.

### 12.5.2 Assisting natural regeneration
The next step in the protection or enhancement of habitat for the benefit of freshwater fishes and other aquatic life is to retain awareness that 'nature knows best' and so to assist natural recovery where ecosystems are damaged or impoverished.

Of all the habitat interventions that can be made for conservation reasons in fresh waters, buffer zones are perhaps the most valuable. A buffer zone is simply a margin of water edge habitat left to grow naturally, be that by fencing out livestock or by leaving an untilled strip between arable land and the water margin. Buffer zones may be simple and cheap, but they enable the protection or regeneration of natural vegetation that reduces direct pollution, provides habitat for all manner of microscopic and larger wildlife of conservation value and also as food and of functional importance, for example, in water purification and flow buffering processes.

A buffer zone for livestock exclusion may comprise a simple strand of barbed wire between posts, but may also be a more elaborate and attractive structure such as a split chestnut fence. Widely spaced wooden fencing may be useful where fields are used for cattle, enabling the animals to graze a limited way beyond the hedge up to the bank top but not to trample this vulnerable zone. Restricted access to the water's edge for drinking purposes can be enabled by a tightly corralled, fenced area, the bed of which may be armoured with gravel. The exact design of a buffer zone is best considered in a very local context, including budget, landowner wishes and additional funding availability, for example, from agri-environment grants. It may need a gate or style to aid access for maintenance or by anglers or for educational purposes.

The 'sell' to the farmer may sometimes be on the basis of a mutual desire to protect aquatic life, as many have an empathy with nature and would like to do the right thing by it though

Buffer zones, segregating tilled or grazed land from the waterfront, allows the regeneration of riparian habitat and attenuates overland flow of sediment and other potential pollutants. (Image © Dr Mark Everard.)

are hampered by the narrow and severe financial constraints of the contemporary farming market. However, co-beneficial issues such as veterinary bills saved by averting hoof disease by animals standing in wet mud, lost time from animals straying into neighbouring land and better control of parasite transmission, as well as avoiding loss of substantial volumes of topsoil may be more persuasive to the landowner. Mutual benefit for fish, other wildlife and farming outcomes are entirely possible from both fenced and untilled buffer strips by the water's edge.

### 12.5.3 Reanimating the water's edge

Sometimes, more active means beyond natural regeneration may be required. For example, a river or pool edge may be recovering from historic straightening, or very severe trampling. In these cases, placing trees and other vegetation in the edge to give nature a 'head start' may be beneficial.

One of the simplest means to achieve this is the 'kneeling' of willows. In simple terms, this entails cutting halfway through the trunk of a waterside willow near the ground and 'hinging' it over into the water. Willows naturally regenerate in water and moist soils, producing a mesh of twigs and new shoots, as well as roots that provide a wealth of food, refuge, spawning and nursery habitat. This is in fact emulating an entirely natural process wherein the trunks and

A 'kneeled willow', in which the main trunk of a small willow or sallow is cut and the bush or tree is 'hinged' into the river margin, provides valuable feeding, refuge and breeding habitat as well as trapping silt and allowing the natural protection of the river bank against erosion. The same effect can be achieved by pegging a fallen willow bough, as in this photograph. The willow continues to grow and put down roots, accreting sediment, stabilising the structure and diversifying river edge habitat still further. (Image © Dr Mark Everard.)

branches of riparian willows naturally fall into the water's edge and root down, armouring and reinforcing banks, beginning the process of river meandering, and thereby enhancing feeding, breeding and refuge habitat.

In other places, a large willow may be cut and lain into the margin, staked with live willow and with a weave of finer willow shoots woven in. At least some of the stakes and weavings generally sprout and regenerate, creating a flow-deflecting structure that reinforces with age, altering flow patterns, sediment deposition and scour and providing a diversity of habitats that a range of water plants will colonise and which will suit the needs of many fish and other wildlife species.

Appropriately designed solutions, commonly requiring permission from regulatory agencies in rivers because of their potential implications for flood management, can greatly improve the functional habitat of impoverished fresh waters, both still and flowing. It is, however, important to let kneeled willows or live willow deflectors establish for two or three years before further management, as overzealous pruning can result in shoots that are not fully established drowning when inundated during flood conditions.

Many other techniques are available to those wanting to improve river habitat, including, for example, protection of eroding banks with staked bundles of woody vegetation that accrete sediment and build up the bank profile as the deposited silt solidifies and plants root in it. But bear in mind that erosion is a natural process, important in rivers for their tendency to meander, so it should not automatically be assumed to be a bad thing.

Where gravels have become silted-up, active measures may be suitable to restore their open pore structure for use by gravel-spawning fish and as a source of food. Gravel jetting entails pumping water through a lance that is plunged into the gravel to dislodge silt, though this is a one-off measure sometimes implemented shortly before these fishes spawn. A more durable method is to stake cut logs into the river bed. These may be a large log perpendicular

Logs staked into a river bed, across the current or as here in an upstream vee, can create turbulence, opening up and desilting gravels, thereby providing spawning habitat in an otherwise silt-laden stream. (Image © Dr Mark Everard.)

to the flow, creating a tumble of water behind it that agitates and opens the gravel, or else an upstream-oriented 'vee' of two logs that has the same net effect.

Warm marginal habitat to stimulate growth and also generate fine food for early life stage fry is also limited in some rivers and pools. One compensatory method here is to open overgrown cattle drinking bays or to dig new scrapes by the water's edge, the shallow and sheltered waters of which warm rapidly on sunny days.

On depleted banks of still waters and rivers, trees can also be actively planted to provide shade, refuge and food. On rivers, headlands on the inside of bends can be fenced off and planted with trees to protect the growing saplings, providing refuge areas between straights. Riparian tree planting or regeneration can be important for many wider reasons, including dissuading larger avian predators like cormorants that tend to fish in more open waters, whilst also providing perches and other useful habitat for other species such as kingfishers that brighten the waterside experience.

### 12.5.4 More active interventions

Many more techniques may be deployed to relieve pressure on fish populations and the wellbeing of other freshwater life, though these tend to be beyond the scope of this book. These include, for example, creating bypass channels around impassable weirs or installation of eel passes through which upstream migrating elvers can gain access through such obstacles to upper rivers. 'Hard' concrete weirs can also be replaced by boulder weirs that are more porous both to water and silt as well as fish and other wildlife.

Other wider opportunities arise from working with farming interests, for example, to apply for grants for a range of river-friendly measures that are also beneficial to farming. For example, these include measures to separate clean roof water from contaminated water in a cattle yard, which not only substantially reduces the volume and likelihood of leakage of 'dirty' water from the yard but can also provide a supply of rainwater to the farmer, averting costs of drawing on piped supplies. Gates can be moved from the bottom of fields uphill and away from water bodies, as concentrated animal and vehicle movement through gates tends to result in unconsolidated mud that can flow with pulses of rainfall into the pool or river. Gate movement also contributes to reducing soil loss and hoof disease and may make vehicle movement easier on higher and drier ground, so being directly beneficial to the farmer.

Britain's pervasive Rivers Trusts are a great source of advice for what can be done to enhance habitat for fishes and other wildlife and are best consulted for expert and context-specific advice.

The chapter 'What have freshwaters fishes ever done for us?' touched on 'fish twitching'. This is the preoccupation of a growing minority (including Mark and Jack!) with an uncommon urge to go out and spot fishes, much as many more people do birds, flowers and butterflies. This is not only good, healthy exercise that also provides informal surveillance of the health of our fresh waters. It is also part of a rising curve of interest in, and recognition of, our rich fish fauna as important and integral elements of healthy ecosystems supporting diverse other organisms and human needs.

## 13.1 FISH TWITCHING AND OTHER FORMS OF ENJOYMENT OF NATURE

Bird 'twitchers' are a far from scarce species in Britain, the million-plus membership of the Royal Society for the Protection of Birds (RSPB) serving as one indicator of the strength of interest in our bird species. Botanising too is a very common informal pastime, hard to quantify but with a massive following and an NGO in the shape of Plantlife to champion Britain's wild flowers, plants and fungi. Butterflies and moths by the water and wider countryside have always enjoyed great popularity. The wider world of invertebrates has not only found renewed life by the inclusion of study of 'mini-beasts' in the primary school curriculum, but the NGO Buglife has increased its profile in public perception and conservation activities, whilst angler-led monitoring of aquatic insects finds a focus in voluntary schemes such as the RiveryFly Partnership. The Earthworm Society of Britain aims to promote and support scientific research about earthworms, their environments and their conservation.

By contrast, there is, as observed in previous chapters, something of a false division in perception of fishes relative to other wildlife. Britain's freshwater fishes may have importance for sport and food, and their distribution may have been significantly rearranged for those

**Aquatic larva of the orange striped stonefly, one of many river fly species indicative of the health of a river system and subject to volunteer river monitoring.**

purposes, but they nevertheless constitute important and functionally significant constituents of our national fauna.

There is now, though, a relatively recent growth in interest in 'fish twitching', peeking into the rich underwater world that has often evaded wider public perception and appreciation. One strand of evidence for this includes the increasing profile of native fishes in television broadcasting. All it takes in reality is a little patience by the waterside, at virtually no expense, for aquatic wonders to reveal themselves.

## 13.2 FISH TWITCHING METHODS

The patient observer, finding a suitable vantage point over still or flowing waters, is likely to be rewarded for their persistence looking through, rather than at, the surface film. This is significantly assisted by a good pair of polarising sunglasses that help vision cut through the surface glare. Keeping still, or walking quietly along the bankside with soft footfalls – remember how sensitive the lateral line of a fish is to vibrations – will enable you to spot any tell-tale movements, either of fish or of changes in surface currents that betray them. Making sure that you have some cover behind you so that you are not a prominent point of the horizon helps you approach fish without them detecting you. Also, avoid both sharp movements or white or otherwise bright clothes that may advertise your presence to fish, which might perceive you as a predator.

Remember to look through the water, maybe focusing on visible areas of river bed or aquatic vegetation below the surface. With luck and patience, your eyes may pick out the gentle swaying of brown trout or grayling over golden gravel as they hold station in a clear water stream. Further downstream, gazing into a swiftly flowing river, one may be rewarded by the dark shapes of barbel or chub ghosting across the river bed between beds of aquatic plants wavering in the current. In summer, the dark forms of chub are readily seen cruising in the warm film of surface water, often willing to accept offerings of bread scraps thrown into the river. You may be lucky enough to spot the elongated profile of a pike, still as death, waiting in ambush in shallower or clear water. At dawn or dusk on a quiet lake, the eye becomes attuned to patches of 'pin bubbles' as tench sift soft sediment for food items down in unseen depths, releasing a fine effervescence of trapped gases through their gills. Occasionally, the wagging tail or an upwelling on the surface may be spied where a common carp is feeding, head down and unleashing a cloud of mud into the water column as it grubs around in a lake bed. At the surface of a summer river, dace and bleak may dimple the surface as they pick at floating insects and other food items. On the lake or in a gravel pit, rudd can often be seen in crimson-finned shoals doing much the same.

Timing may be important, as our fish species are often most active at dawn and dusk. As roach and common bream come out to feed at dusk, they can be seen 'priming' as they

A brown trout idling in the current below a bridge, a nice sight to reward the attentions of a keen 'fish twitcher'.

swim to the surface to gulp in air, recharging their swim bladders to change their buoyancy to accommodate sorties into shallower water. Maybe you will spot a surface 'explosion' as a shoal of small fry scatters in frenzy as a posse of hungry perch strikes at them from beneath.

A couple of slices of bread in the pocket, or some floating dog biscuits, might encourage a common carp to feed at the surface, and you may hear and then subsequently tune into the greedy 'clooping' noise it makes as it sucks down your free offering. Chub too are often readily brought to the surface to feed, as may be rudd in still water, and bleak and dace in flowing water.

Some of the more rewarding sights for the fish twitcher are to see an Atlantic salmon leaping over a weir or other obstruction as the autumnal upstream migration gets under way from October onwards, though this may take a little more preparation to put you in the right place at the right time.

For the more adventurous fish twitcher, a small and cheap aquarium net may expose some of the smaller wonders in the margins of freshwater bodies. So too may one of the increasingly available and affordable submersible cameras. The intrepid may choose to truly immerse themselves with a snorkel and mask.

But no special access or equipment is required for the enjoyment of the nascent fish twitcher. They will be quickly rewarded with a little diligence, and perhaps the company of a more experienced companion, by new sights and experiences of our wonderful freshwater environments including their formerly unsuspected and underappreciated fishy fauna.

An Atlantic salmon leaping a weir or other obstruction is a rewarding sight for the fish twitcher. (Image credit: Kevin Wells Photography/Shutterstock.com.)

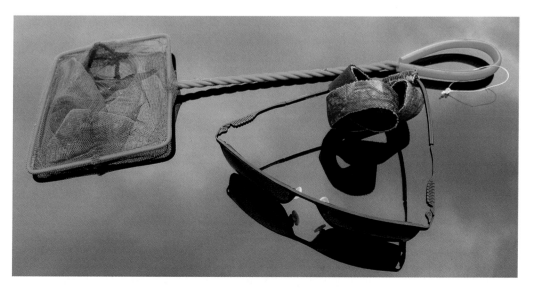

A good pair of polarising sunglasses will help you see through the surface glare to what lies beneath, and the more intrepid twitcher may brandish an aquarium net in the water margin to see what small life is to be found there. (Image © Dr Mark Everard.)

The most important considerations are to keep safe, not putting yourself at any risk, and to enjoy the experience!

## 13.3 FISH CARE WHEN FISH TWITCHING

Generally, fish twitching is non-invasive, causing little or no disturbance. However, some precautions are necessary to ensure that the fish come to no harm.

Firstly, be aware when fish are spawning, as they are often at their most conspicuous and vulnerable. This is a good time to see tench and carp cavorting in lake vegetation, roach and bream in shoals rolling in near-bank surfaces including those covered in mosses, other water plants or against reed stems or salmon, trout, chub and barbel in their season splashing on gravel redds. But you may inadvertently scare them off prime habitat if you approach too closely.

The second important consideration is biosecurity, addressed in the previous section. Wash and dry boots, nets and other equipment in case some unwanted stowaway organism accompanies you to the waterside.

Thirdly, do take account of the needs of the fishes that interest you. For example, it is vital to respect the fact that bullheads are loyal to stones or other caves under woody debris and other structures. An adult bullhead can live out its whole life under a single rock, as a source of shelter, respite from strong currents, opportunities to feed on the invertebrates colonising the rock or settling in the slack water around it and a place to nest and nurture young. So, if you ever set out to turn over stones to find bullheads, or to catch them with your bare hands, do always remember to turn the stones back and return the fish exactly where you found it. Otherwise, the bullhead may be disoriented or, even worse, separated from a brood of eggs.

**If you turn stones to find bullhead, always put the stone back exactly as you found it as bullheads are highly territorial and may be disoriented if their home cave is not replaced exactly.**

**Some locations are popular with visitors who can see trout feeding on free offerings.**

## 13.4 FISH TWITCHING TOURISM

In India, many temples have sacred pools of fish that may also be a tourism draw. Some pubs in Britain, for example, the Salmon Leap Pub near Southampton, trade on the fact that patrons stand a good chance of seeing salmon leaping an adjacent weir in the late autumn migratory season. Other towns, such as Stockbridge on the River Test, can be visited in the confident expectation of seeing large trout idling in the current beneath bridges and eager to take any proffered scrap of bread, biscuit or chip. The Derbyshire Wye running through the town of Bakewell is also another popular spot where the public come to see trout feeding on offerings such as bread, chips and, of course, fragments of Bakewell tart! Less prominently, but equally part of the experience, other towns with river crossings and nature reserves with water bodies include sightings of fish as part of the day's experience.

Fish do in some places, and can in others, form part of the tourism experience – for the casual or committed fish twitcher even if they have no angling interest – contributing to the tourism experience and economy.

## 13.5 FISH TWITCHING FOR FUN

With appropriate care and a little patience, and ideally guidance from a more experienced friend or colleague, I can strongly recommend the simple, almost childish enjoyment of fish twitching. It is a healthy, fun way to appreciate nature and to help us realise more deeply its wonder and the need to protect it better from the multiple pressures of modern society as a fundamental contribution to our aspirations for a sustainable and enjoyable future.

# SO WHAT HAVE WE LEARNED ABOUT BRITAIN'S FRESHWATER FISHES?

One of the seminal books at the outset of the contemporary environmental movement was Rachel Carson's 1962 work *Silent Spring*. The book was excellent in collating in accessible terms the body of knowledge emerging globally about the accumulation of persistent pesticides in nature, even in some of the most remote quarters of the globe where they could not have been released directly. *Silent Spring* drew a dire yet well-founded prognosis for the impacts on wildlife and human health. Adding considerably to the book's impact was its concise and evocative title, forewarning of a spring bereft of birdsong.

Woodland and other semi-natural habitats devoid of birds, or for that matter butterflies, bees and blossoms, is something we all feel to be wrong regardless of the depth of our ecological knowledge. The same is true of a fishless river, canal, pond or lake. The nuances of species, communities and age profiles within them, and distinctions between native and introduced species, may escape many people. However, the presence or absence of fish nevertheless has meaning for the lay public and sometimes, also, more specific spiritual and cultural meanings.

Better understanding of our fish fauna is therefore important, in terms of the ways they enrich our appreciation of the natural world but also in our ability to value and protect it. So it is important that we overcome a historic perception of separateness between freshwater fishes and other constituents of British wildlife, a message repeated frequently throughout this book. We have to learn to appreciate how integrally embedded and co-evolved fishes are in ecosystems, the often significant roles they play in the diverse processes that maintain ecological diversity and functioning and the numerous benefits that the natural world confers upon humanity.

**Fish are integral elements of freshwater ecosystems and need to be respected and protected as such.**

At the time of writing, the Royal Society for the Protection of Birds (RSPB) has a membership well in excess of one million. Birds are far more visible than life under water, so perhaps an aspiration to set up an equivalent fish-oriented body attracting such a large number of subscribers is fanciful. Nevertheless, what lives below the waterline is a very important and often neglected part of our wildlife, influencing the character and utility of water bodies but with far wider ramifications across terrestrial catchments and coastal ecosystems.

There has also been a slow but cumulatively substantial shift in paradigm of fishery management from the latter half of the twentieth century. We are progressively moving away from fixing the symptoms of low perceived fish stocks, for example, by stocking and predator control, towards a far more ecosystem-centred approach that seeks to rebuild the resilience of fish stocks through techniques such as habitat enhancement for spawning and recruitment, refuge and food availability. This has benefits for the whole aquatic ecosystem and its diverse values to society. We need to see a wider recognition, respect and inclusion of our precious and diverse freshwater fish fauna in wider conservation, environmental management and education initiatives. Greater outcomes for both nature and people ensue when the ways we manage all of our ecosystems, including those perceived or used as fisheries, take an holistic approach. This is irrespective of whether the primary focus of attention is upon mammals, birds, reptiles and amphibians, fish, fungi, plants or other wildlife, or protection of water quality, natural approaches to flood management, landscape aesthetics or other purposes, including managing exploitable fish populations. All of these end points are intimately connected, winning or losing together, and great synergies and cost savings are possible if we work together across the many artificial barriers we erect between sectoral interests.

Fish are far more than something enjoyed in batter or observed from a river bridge or swimming in a tank in a dentist's waiting room. To ask a somewhat surreal question, what is the meaning of fish? The answers to this question are as diverse as the span of human perceptions and value systems, from the utilitarian and commercial to the cultural, aesthetic, spiritual, recreational, functional and inherent.

If this book has served to raise your appreciation of our wonderful and diverse wealth of freshwater fishes and to rehabilitate their place amongst the pantheon of British wildlife, it will have served valuable purposes. If you have enjoyed the read, I am happy too.

# *Further reading*

Many books, articles and scientific papers have been written about Britain's freshwater fishes, or at least aspects of them. Just a few relevant ones are listed here of general interest or supporting topics addressed in a little more detail in this book. These publications are broken down into 'General books about fishes', 'Books about specific fish species or groups of fishes', 'Books relating to conservation and management' and 'Other general or special interest books, papers and reports', many of which are cited or support parts of this text.

## GENERAL BOOKS ABOUT FISHES

Buckland, F. (1880). *Natural History of British Fishes*.

Cholmondeley-Pennell, H. (1863). *The Angler Naturalist*.

Everard, M. (1994). Rivers and Lakes. In: *Nature Worlds*. Duncan Baird Publishers, London, pp. 106–113.

Everard, M. (2008). *The Little Book of Little Fishes*. Medlar Press, Ellesmere, 192 pp.

Everard, M. (2012). *Fantastic Fishes: A Feast of Facts and Fables*. Medlar Press, Ellesmere.

Everard, M. (2013). *Britain's Freshwater Fishes*. Princeton University Press/WildGUIDES.

Everard, M. (2016). *Know your Freshwater Fishes*. Old Pond Publishing, Sheffield.

Giles, N. (1993). *Freshwater Fish of the British Isles*. Swan Hill Press, Shrewsbury.

Houghton, R.W. (1879). *British Fresh-Water Fishes*.

Kottelat, M. and Freyhof, J. (2007). *Handbook of European Freshwater Fishes*. Published by the authors.

Maitland, P. (2004). *Key to the Freshwater Fish of Britain and Ireland, with Notes on their Distribution and Ecology*. Freshwater Biological Association Scientific Publications 62. Freshwater Biological Association, Ambleside.

Mansfield, K. (1958). *Small Fry and Bait Fishes: How to Catch Them*. Herbert Jenkins, London.

Muus, B.J. and Dahlstrom, P. (1972). *Collins Guide to the Freshwater Fishes of Britain and Europe*. Collins, Glasgow.

Salter, T.F. (1815). *The Angler's Guide*.

Wheeler, A. (1969). *The Fishes of the British Isles and North West Europe*. Macmillan and Co., London.

Williamson, C.T. (1808). *The Complete Angler's Vade-Mecum*.

## BOOKS ABOUT SPECIFIC FISH SPECIES OR GROUPS OF FISHES

Everard, M. (2006). *The Complete Book of the Roach*. Medlar Press, Ellesmere, 436 pp.

Everard, M. (2008). *The Little Book of Little Fishes*. Medlar Press, Ellesmere, 192 pp.

Everard, M. (2009). *Barbel River*. Medlar Press, Ellesmere, 126 pp.

Everard, M. (2011). *Dace: The Prince of the Stream*. MPress, Romford, 248 pp.

Everard, M. (2013). *Redfin Diaries: A Life in the Year of a Roach Enthusiast*. Coch-y-Bonddu Books, Machynlleth.

Everard, M. and Knight, P. (2013). *Britain's Game Fishes: Celebration and Conservation of Salmonids*. Pelagic Press, Totnes.

Pinder, A.C. (2001). *Keys to Larval and Juvenile Stages of Coarse Fishes from Freshwaters in the British Isles*. Freshwater Biological Association Scientific Publication No.60, Freshwater Biological Association, Windermere.

## BOOKS RELATING TO CONSERVATION AND MANAGEMENT

Everard, M. (2015). *River Habitats for Coarse Fish: How Fish use Rivers and how we can Help Them*. Old Pond Publishing, Sheffield.

Wild Trout Trust. (2017). *The Wild Trout Survival Guide, 4th edition*. The Wild Trout Trust.

## OTHER GENERAL OR SPECIAL INTEREST BOOKS, PAPERS AND REPORTS

Bacon, F. (1623). *History of life and death, of the second title in natural and experimental history for the foundation of philosophy being the third part of the instauratio magna*.

Bailey, J.M. (1997). *Tales from the Riverbank*. BBC Books, London.

Banis, V.J. (2012). *Charms, Spells, and Curses*. eBook published by Wildside Press.

Beville S. (2005). *The economic significance of the fisheries of the Test and Itchen*. The Test and Itchen Association Ltd Rivers Report, pp. 20–23.

Blackstone, Sir W. (1765–1769). *Commentaries on the Laws of England*. Clarendon Press, Oxford.

Brown, A., Stolk, P. and Djohari, N. (2010). Angling: A social research overview. Substance (a report commissioned by the Environment Agency). http://www.resources.anglingresearch.org.uk/sites/resources.anglingresearch.org.uk/files/Substance-EA-Angling-Research-Review.pdf

Carty, P. and Payne, S. (1998). *Angling and the Law*. Merlin Unwin Books, Ludlow.

Currie, C.K. (1991). The early history of the carp and its economic significance. *The Agricultural History Review*, 39(22), pp. 97–107.

Department of Health. (2004). *Choosing Health (White Paper)*. Department of Health, London.

Environment Agency. (2003). *National Trout and Grayling Fisheries Strategy*. Environment Agency, Bristol.

Everard, M. (2010). Ecosystem services assessment of buffer zone installation on the upper Bristol Avon, Wiltshire. Environment Agency Evidence Report SCHO0210BRXW-e-e.pdf. Environment Agency, Bristol.

Everard, M. and Jevons, S. (2010). Ecosystem services assessment of sea trout restoration work on the River Glaven, North Norfolk. Environment Agency Evidence Report SCHO0110BRTZ-e-e. Environment Agency, Bristol.

Everard, M., Pinder, A.C., Raghavan, R. and Kataria, G. (2019). Viewpoint: Are well-intended Buddhist practices an under-appreciated threat to global aquatic biodiversity? *Aquatic Conservation: Marine and Freshwater Ecosystems*, 29(1), pp. 136–141.

Harwood, K. (2017). *Potted char*. Waterlog No.100, Summer 2017.

Le Cren, D. (2001). The Windermere perch and pike project: A historical review. *Freshwater Forum*, 15, pp. 3–34.

Nash, C. (2011). *The History of Aquaculture*. Wiley-Blackwell.

Neville, M. (2009). Fishy tale of a real gold fish. *Isle of Wight County Press*, Friday 1 May 2009. http:// www.iwcp.co.uk/news/news/fishy-tale-of-a-real-gold-fish-25933.aspx, accessed 29th January 2011

Ricardi, A. and Rasmussen, J.B. (1998). Perspectives: Predicting the identity and impact of future biological invaders: A priority for aquatic resource management. *Canadian Journal of Fisheries and Aquatic Sciences*, 55, pp. 1759–1765.

Stewart, C. (1817). *Elements of the Natural History of the Animal Kingdom: Comprising the Characters of the whole Genera, and of the Most Remarkable Species, Particularly those that are Natives of Britain, Volume 1*. Bell and Bradfute, Edinburgh.

Substance. (2019). Angling for good: National angling strategy 2019–2024. Substance (in collaboration with the Environment Agency, Angling Trust, Canal and River Trust, Angling Trade Association, Get Hooked on Fishing, and Sport England). https://www.substance.net/wp-content/uploads/AnglingforGood.pdf

The Age. (2008). Schoolboy explodes goldfish memory myth. The Age, 18th February 2008. http://www.theage.com.au/articles/2008/02/18/1203190696599.html

The Telegraph. (2017). Swedish eel slithers its last after 155 years: Eel thought to be the world's oldest found dead in family well. *The Telegraph*, 6 November 2017. http://www.telegraph.co.uk/news/worldnews/europe/sweden/11023470/Swedish-eel-slithers-its-last-after-155-years.html

UN Food and Agriculture Organization (FAO). Yearbooks 1996 to 2005, Fishery Statistics, Commodities. Volumes 83–97. FAO: Rome, Italy.

Walton, I. (1653). *The Compleat Angler*.

Ward, A.M. (1974). Crassus' slippery eel. *The Classical Review*, 24, pp. 185–186.

Yarrell, W. (1836). *The History of British Fishes*. John Van Voorst, London.

Yasuo, W. (2004). Venomous freshwater fish. *Japanese Journal of Clinical Dermatology*, 58(5), pp. 11–14.

Zhu, H.P., Ma, D.M. and Gui, J.F. (2006). Triploid origin of the gibel carp as revealed by 5S rDNA localization and chromosome painting. *Chromosome Research*, 14(7), pp. 767–776. DOI: 10.1007/s10577-006-1083-0.

# Appendix 1: Glossary

**Adipose fin:** A small, soft and fleshy fin composed of fatty tissue, located between the dorsal and caudal (tail) fin.

**Alevin:** The first larval life stage of salmon and trout to emerge from the egg, the larva still largely undeveloped and attached to a yolk sac.

**Allochthonous:** An aquatic ecosystem in which most of the food arrives from outside the water body.

**Ammocoete:** The freshwater, larval life stage of lamprey species.

**Anadromous:** Fish that live their adult lives in the sea or estuaries but return to fresh waters to breed.

**Anal fin:** A fin located on the lower surface of the body of a fish, between the anus and the caudal fin. In some fishes, such as eels and lampreys, the anal fin is contiguous with the caudal fin. The anal fin stabilises the fish whilst swimming.

**Autochthonous:** An aquatic ecosystem in which most of the food is produced within the water body.

**BAP:** Also UK BAP: the UK Biodiversity Action Plan.

**Barbels:** Sometimes referred to as 'whiskers', barbels are slender and tactile, whisker-like organs generally densely covered with chemical sense organs around the mouth and/ or on the snout of some fishes.

**Baseflow:** Water entering rivers and standing waters via groundwater.

**Baseflow index:** The amount of groundwater entering rivers over a long-term average as a proportion of total river flow.

**BFI:** See 'baseflow index'.

**Bioaccumulative:** A chemical tending to accumulate in biological systems.

**Biodiversity Action Plan:** See BAP.

**Benthic:** The benthic zone relates to the region near and at the bed of a water body. Animals and plants living here are known as benthic organisms.

**Brackish:** Saline water generally found in estuaries that is not as fully saline as sea water.

**Carnivorous:** A diet that comprises mainly animals.

**Catadromous:** Fishes that live their adult lives in fresh waters but return to the sea to breed.

**Catch-and-release:** A conservation practice in recreational angling in which fish are unhooked once caught and returned alive to the water.

**Catchment:** The geographical area from which water flows into a river system, lake or other water body.

**Caudal fin:** Also known as the tail fin, a fin at the rear of the fish.

**Caudal peduncle:** See *Peduncle*.

**Closed season:** A statutory period during which angling and/or commercial fishing is banned for stock protection purposes, generally during spawning periods.

**Coarse fish:** In a British freshwater context, a fish that is not a member of the salmon family, though generally including the grayling which spawns during the spring like most non-salmonid species.

**Crepuscular:** Active at dusk and dawn.

**Density dependence:** Limitation on population density due to the availability of suitable habitat and the territorial behaviour of the species, for example of Atlantic salmon parr competing for suitable riffle habitat.

**Detritus:** Amorphous organic matter that may be abundant in slow-moving and still water bodies, formed from decomposition of plant matter and faecal matter and potentially rich in microorganisms. Detritus may be consumed by omnivorous fishes when other food is scarce.

**Dorsal fin:** Single or multiple fins on the back of the fish.

**Filter-feeding:** The habit of straining suspended food particles from water, typically in fish by passing the water over modified gill rakers.

**Fins:** Flattened surfaces used by fish for stability and/or to produce lift, thrust or steerage in water. In bony fish, most fins are held erect by soft rays and often also spines.

**Flashiness:** The tendency of a river or stream to respond rapidly to rainfall, rising and dropping again rapidly where flow buffering is low.

**Fry:** Early juvenile, free-swimming life stages of fishes, from the point they become reliant upon external sources for nutrition.

**Game fish:** In a British freshwater context, a fish that is a member of the salmon family, though generally excluding the grayling which spawns during the spring like most non-salmonid species.

**Gill(s):** The respiratory organs of fish, allowing the inward diffusion of dissolved oxygen and outward diffusion of carbon dioxide and other waste matter between water and blood. Gills consist of thin filaments of tissue with a highly folded surface through which fine blood vessels flow, arranged on gill arches supporting them in a flow of water at the back of the mouth. The vulnerability of the gills means that they are armoured with gill covers, which also assist with respiration.

**Gill arches:** Also known as branchial arches, a series of bony structures beneath the gill covers of fish that support the gill filaments.

**Gill cover:** Also known as the operculum, a hard bony flap covering and protecting the gills in most fishes. The gill cover comprises four fused bones, apparently evolved by the joining of gill-slit covers of ancestor species.

**Gill rakers:** Filamentous bony or cartilaginous, forward-facing extensions from the gills, generally used for filtering fine food particles from the water.

**Glacial relics:** Fish adapted to cold waters, often found in as scattered populations in deep lakes, presumed to be relics of a colder glacial period of history.

**Guanine:** A pigment that imparts the reflective, silvery colour over the body surface of many fish species.

**Holarctic:** The northern regions of America, Greenland and Eurasia.

**Ichthyology:** The branch of the scientific study of animals (zoology) devoted to the fishes.

**Inferior:** Relating to the position of the mouth, one that is underslung or oriented downwards.

**Invasive:** A species introduced beyond its native range that can establish self-sustaining populations potentially threatening the balance of ecosystems.

**Kype:** The hooked lower jaw that develops in mature salmonid fishes prior to spawning, used as a weapon to defend spawning territories and potential mates against other competing male fish.

**Lateral line:** A chain of pressure sensory organs along the flanks of most fish species, helping them to detect movement, vibration and pressure waves in the surrounding water. The number of scales along the lateral line may be useful in identification of some species.

**Maxillary:** Relating to a jaw or jawbone, especially the upper one.

**Milt:** Seminal fluid of fish containing sperm and nutritious substances, released into the water to fertilise eggs laid by female fish.

**Minor species:** A somewhat dismissive term used to group smaller fish species that are nevertheless important members of aquatic ecosystems, some of them of priority conservation importance.

**Natal:** The place something is born or, in the case of British freshwater fish species, hatched.

**Naturalised:** A species introduced beyond its native range that has become established as a viable breeding population.

**Nursery habitat:** An area of habitat suiting the needs of juvenile fish, providing refuge from predation and strong flows as well as abundant suitable small food items.

**Oligotrophic:** A naturally nutrient-poor water.

**Omnivorous:** Eats a variety of plants, animals and amorphous organic matter.

**Operculum:** See gill cover.

**Overbite:** A mouth structure in which the upper jaw extends beyond and over the lower jaw.

**Ovipositor:** A tube growing from the vent of some female fish (bitterling in breeding condition) through which they deposit eggs.

**Panning:** Compression of the soil surface, often by agricultural activities, that creates a surface layer impenetrable to moisture.

**Parr:** The longest-lasting freshwater stage of a trout or migratory salmon, between the fry stage and the smolt.

**Pectoral fins:** Paired fins located on each side of the body, generally just behind the gill covers, used for 'paddling' and delicate orientation, as well as for lift and maintenance of depth in currents.

**Peduncle:** The caudal peduncle of many species of fish is the narrow part of the body, generally tapering in width, to which the tail fin attaches.

**Pelagic:** The pelagic zone is open water that is neither close to the bottom nor near the shore. Animals and plants living here are known as pelagic organisms.

**Pelvic fins:** See ventral fins.

**Pharyngeal teeth:** Also known as throat teeth, teeth found in the throat of fishes of the carp and minnow family (*Cyprinidae*) that lack teeth in the jaw or mouth. Pharyngeal teeth are located in the throat, modified from gill arches and are used to crush or grind hard food items.

**Piscivorous:** The habit of eating fish.

**Planktivorous:** The habit of feeding on plankton or small organisms suspended in the water column.

**Precocious parr:** In salmon, a proportion of parr that mature sexually before migrating to sea and that may fertilise eggs released by spawning females.

**Priming:** The behaviour of breaking the water's surface to gulp in air to charge the swim bladder, seen in cyprinids (the carp and minnow family, *Cyprinidae*).

**Rays:** Soft, branching structures supporting the fins.

**Recruitment:** Addition of new juvenile fish to the population.

**Redds:** Depressions cut into the gravel of river or lake beds generally by female fish of the salmon family or by lampreys to deposit eggs.

**Runoff:** Water reaching rivers and standing waters by overground flow.

**Scales:** Hardened bony plates varying in size, shape, structure and extent as an armour on the surface of a fish; also a characteristic feature used to identify fish families and species.

**Smolt:** The phase of a salmon or migratory trout's life in which it takes on a silver sheen, migrates down the river and metamorphoses into the sea-going adult form.

**Spate:** A surge of water down a river or stream, generally in response to a significant rainfall event.

**Spigin:** Glue protein produced from the kidney of male stickleback used in building nests from pieces of vegetation.

**Spine:** Stiff and sharp projections stiffening fins, often just the leading edge in front of branched rays. Some species such as perch, bullhead and European seabass also have spines on the rear of the gill covers.

**Springer:** Shorthand for a spring salmon, one that returns from the sea to run rivers early in the year. Springers are often larger specimens.

**Superior:** Relating to the position of the mouth, one that is oriented upwards.

**Swim bladder:** Also known as the gas bladder or air bladder, an internal gas-filled sac enabling fish to control their buoyancy and increasing the ability of the fish to detect sounds.

**Tail fin:** See caudal fin.

**Tapetum lucidum:** A reflective layer behind the retina aiding hunting in low light.

**Tubercles:** Small, horny swellings that develop on many fish species, particularly males in the carp and minnow family (*Cyprinidae*), as a prelude to spawning. They develop most densely around the head and front of the body and are believed to provide tactile stimuli that promote the release of eggs by females and aid the male's adhesion to the female's flank when releasing their milt.

**Ventral fins:** Also known as pelvic fins, ventral fins are paired and located ventrally behind the pectoral fins but before the anus. Ventral fins are vestigial in the European eel and are missing entirely in the lampreys.

**Vomerine teeth:** Small patches of teeth found on the front part of the roof of the mouth in some species, such as pike, perch, zander and European seabass.

**Zooplankton:** Small animals suspended in the water column.

# Appendix 2: Distribution maps of British freshwater fishes

Distribution maps were supplied by the Biological Records Centre (BRC) on behalf of the Freshwater Fish Recording Scheme. BRC receive support from the Joint Nature Conservation Committee and the UK Centre for Ecology & Hydrology (via the Natural Environment Research Council award number NE/R016429/1 as part of the UK-SCAPE programme delivering National Capability). We thank Ian Winfield who coordinates the Freshwater Fish Recording Scheme and Steph Rorke (BRC) for supplying maps. We are indebted to volunteer recorders and organisations who provide data to the scheme, including a substantial contribution of records from the Environment Agency.

In the following maps, different coloured dots demonstrate date ranges from which records derive, which is of interest in terms of demonstrating how the ranges of some species have spread over time. Although distribution maps are available for hybrids, they are not included in this Appendix.

# A2.1 THE LEUCISCINE, OR DACE-LIKE, CYPRINID FISHES (CARP AND MINNOW FAMILY)

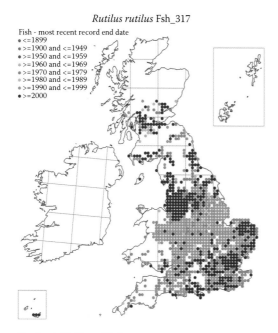

*Rutilus rutilus* Fsh_317

Fish - most recent record end date
- <=1899
- >=1900 and <=1949
- >=1950 and <=1959
- >=1960 and <=1969
- >=1970 and <=1979
- >=1980 and <=1989
- >=1990 and <=1999
- >=2000

The Roach (*Rutilus rutilius*)

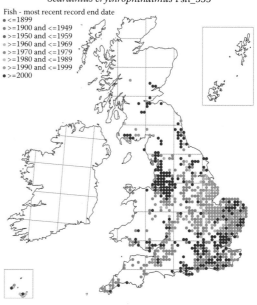

*Scardinius erythrophthalmus* Fsh_335

Fish - most recent record end date
- <=1899
- >=1900 and <=1949
- >=1950 and <=1959
- >=1960 and <=1969
- >=1970 and <=1979
- >=1980 and <=1989
- >=1990 and <=1999
- >=2000

The Rudd (*Scardinius erythrophthalmus*)

*Leuciscus cephalus* Fsh_254

Fish - most recent record end date
- <=1899
- >=1900 and <=1949
- >=1950 and <=1959
- >=1960 and <=1969
- >=1970 and <=1979
- >=1980 and <=1989
- >=1990 and <=1999
- >=2000

The Chub (*Squalius cephalus*)

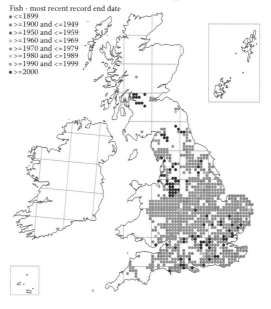

*Leuciscus leuciscus* Fsh_258

Fish - most recent record end date
- <=1899
- >=1900 and <=1949
- >=1950 and <=1959
- >=1960 and <=1969
- >=1970 and <=1979
- >=1980 and <=1989
- >=1990 and <=1999
- >=2000

The Dace (*Leuciscus leusciscus*)

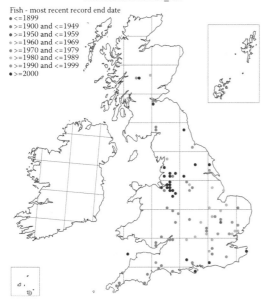

The Orfe or Ide (*Leuciscus idus*)

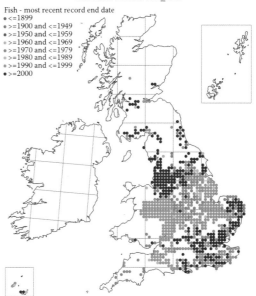

The Common or Bronze Bream (*Abramis brama*)

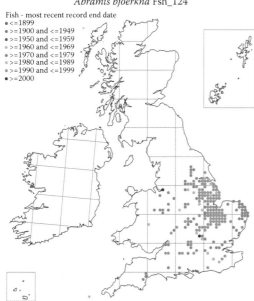

The Silver Bream (*Blicca bjoerkna*)

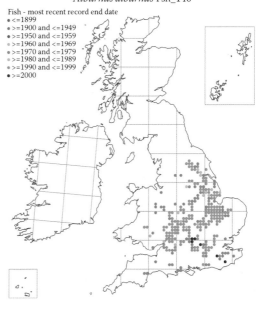

The Bleak (*Alburnus alburnus*)

# A2.2  OTHER MEMBERS OF THE CYPRINID (CARP AND MINNOW) FAMILY

### *Cyprinus carpio* Fsh_215

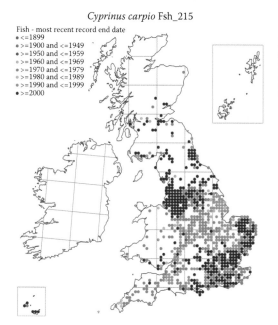

The Common Carp (*Cyprinus carpio*)

### *Carassius carassius* Fsh_183

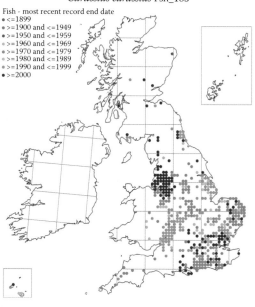

The Crucian Carp (*Carassius carassius*)

### *Carassius auratus* Fsh_180

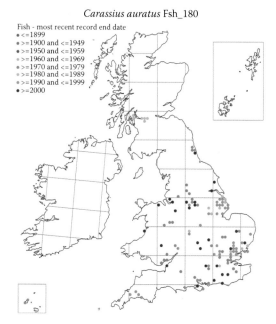

The Goldfish (*Carassius auratus*)

### *Ctenopharyngodon idellus* Fsh_210

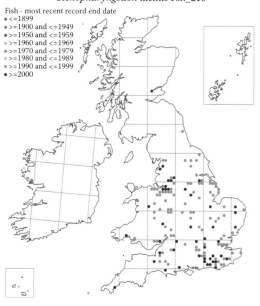

The Grass Carp (*Ctenopharyngodon idella*)

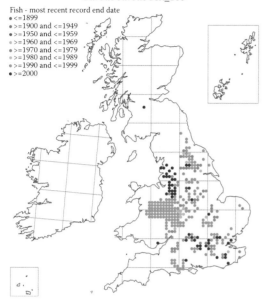

**The Barbel (*Barbus barbus*)**

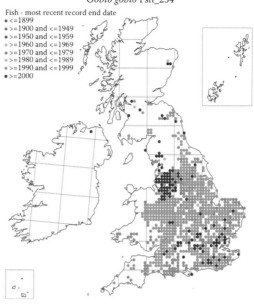

**The Gudgeon (*Gobio gobio*)**

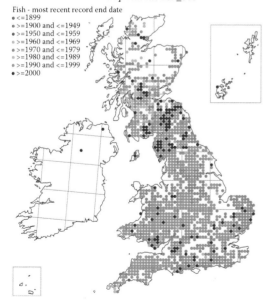

**The Minnow (*Phoxinus phoxinus*)**

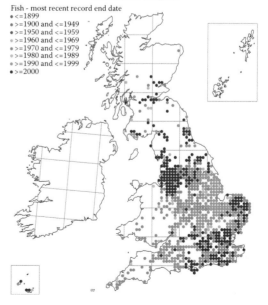

**The Tench (*Tinca tinca*)**

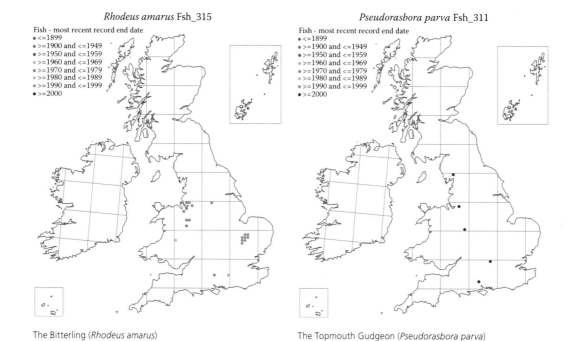

### Rhodeus amarus Fsh_315

Fish - most recent record end date
- <=1899
- >=1900 and <=1949
- >=1950 and <=1959
- >=1960 and <=1969
- >=1970 and <=1979
- >=1980 and <=1989
- >=1990 and <=1999
- >=2000

The Bitterling (*Rhodeus amarus*)

### Pseudorasbora parva Fsh_311

Fish - most recent record end date
- <=1899
- >=1900 and <=1949
- >=1950 and <=1959
- >=1960 and <=1969
- >=1970 and <=1979
- >=1980 and <=1989
- >=1990 and <=1999
- >=2000

The Topmouth Gudgeon (*Pseudorasbora parva*)

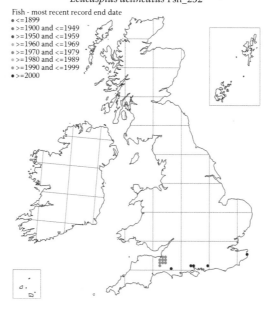

### Leucaspius delineatus Fsh_252

Fish - most recent record end date
- <=1899
- >=1900 and <=1949
- >=1950 and <=1959
- >=1960 and <=1969
- >=1970 and <=1979
- >=1980 and <=1989
- >=1990 and <=1999
- >=2000

The Sunbleak (*Leucaspius delineatus*)

## A2.3 FISHES OF THE SALMON FAMILY

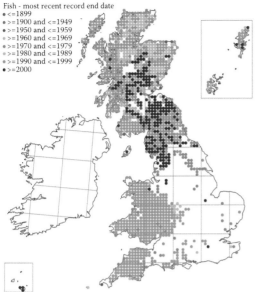

*Salmo salar* Fsh_322

Fish - most recent record end date
- <=1899
- >=1900 and <=1949
- >=1950 and <=1959
- >=1960 and <=1969
- >=1970 and <=1979
- >=1980 and <=1989
- >=1990 and <=1999
- >=2000

The Atlantic Salmon (*Salmo salar*)

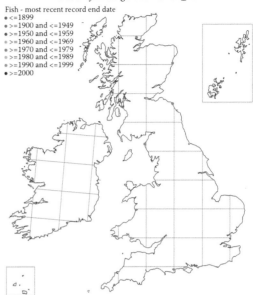

*Oncorhynchus gorbuscha* Fsh_289

Fish - most recent record end date
- <=1899
- >=1900 and <=1949
- >=1950 and <=1959
- >=1960 and <=1969
- >=1970 and <=1979
- >=1980 and <=1989
- >=1990 and <=1999
- >=2000

The Pink Salmon (*Oncorhynchus gorbuscha*)

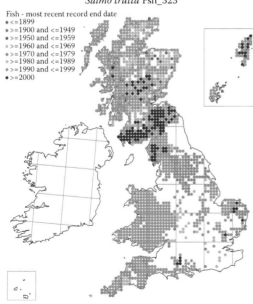

*Salmo trutta* Fsh_323

Fish - most recent record end date
- <=1899
- >=1900 and <=1949
- >=1950 and <=1959
- >=1960 and <=1969
- >=1970 and <=1979
- >=1980 and <=1989
- >=1990 and <=1999
- >=2000

The Brown Trout or Sea Trout (*Salmo trutta*)

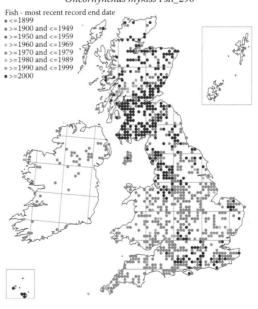

*Oncorhynchus mykiss* Fsh_290

Fish - most recent record end date
- <=1899
- >=1900 and <=1949
- >=1950 and <=1959
- >=1960 and <=1969
- >=1970 and <=1979
- >=1980 and <=1989
- >=1990 and <=1999
- >=2000

The Rainbow Trout (*Oncorhynchus mykiss*)

## Salvelinus fontinalis Fsh_332

Fish - most recent record end date
- ≤1899
- ≥1900 and ≤1949
- ≥1950 and ≤1959
- ≥1960 and ≤1969
- ≥1970 and ≤1979
- ≥1980 and ≤1989
- ≥1990 and ≤1999
- ≥2000

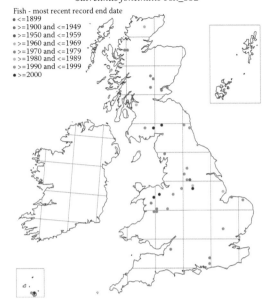

The Brook Trout (*Salvelinus fontinalis*)

## Thymallus thymallus Fsh_357

Fish - most recent record end date
- ≤1899
- ≥1900 and ≤1949
- ≥1950 and ≤1959
- ≥1960 and ≤1969
- ≥1970 and ≤1979
- ≥1980 and ≤1989
- ≥1990 and ≤1999
- ≥2000

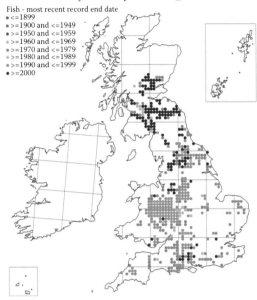

The Grayling (*Thymallus thymallus*)

## Salvelinus alpinus Fsh_331

Fish - most recent record end date
- ≤1899
- ≥1900 and ≤1949
- ≥1950 and ≤1959
- ≥1960 and ≤1969
- ≥1970 and ≤1979
- ≥1980 and ≤1989
- ≥1990 and ≤1999
- ≥2000

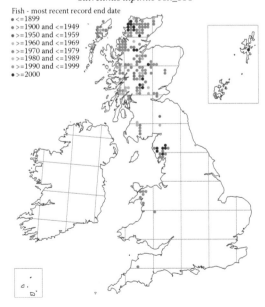

The Alpine Charr (*Salvelinus alpinus*)

## Coregonus lavaretus Fsh_204

Fish - most recent record end date
- ≤1899
- ≥1900 and ≤1949
- ≥1950 and ≤1959
- ≥1960 and ≤1969
- ≥1970 and ≤1979
- ≥1980 and ≤1989
- ≥1990 and ≤1999
- ≥2000

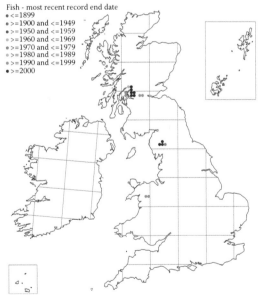

The European Whitefish (*Coregonus lavaretus*)

*Coregonus albula* Fsh_202

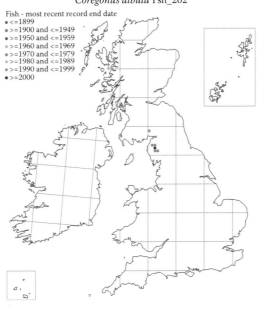

The Vendace (*Coregonus albula*)

## A2.4 THE PIKE

*Esox lucius* Fsh_222

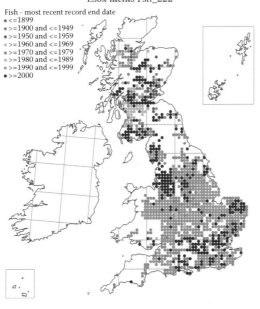

The Pike (*Esox lucius*)

## A2.5 FISHES OF THE PERCH FAMILY

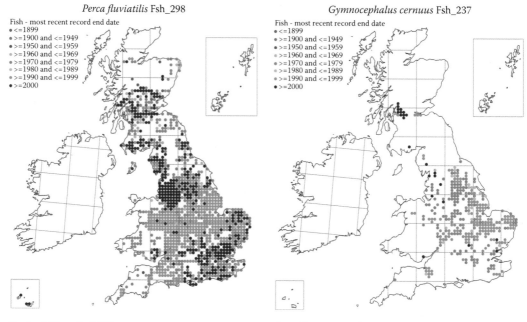

### Perca fluviatilis Fsh_298

Fish - most recent record end date
- <=1899
- >=1900 and <=1949
- >=1950 and <=1959
- >=1960 and <=1969
- >=1970 and <=1979
- >=1980 and <=1989
- >=1990 and <=1999
- >=2000

The Perch (*Perca fluviatilis*)

### Gymnocephalus cernuus Fsh_237

Fish - most recent record end date
- <=1899
- >=1900 and <=1949
- >=1950 and <=1959
- >=1960 and <=1969
- >=1970 and <=1979
- >=1980 and <=1989
- >=1990 and <=1999
- >=2000

The Ruffe (*Gymnocephalus cernua*)

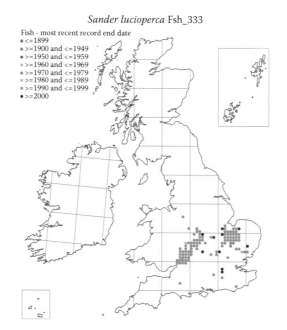

### Sander lucioperca Fsh_333

Fish - most recent record end date
- <=1899
- >=1900 and <=1949
- >=1950 and <=1959
- >=1960 and <=1969
- >=1970 and <=1979
- >=1980 and <=1989
- >=1990 and <=1999
- >=2000

The Zander (*Sander lucioperca*)

# A2.6 OTHER SMALLER FISHES

### *Gasterosteus aculeatus* Fsh_229

Fish - most recent record end date
- <=1899
- >=1900 and <=1949
- >=1950 and <=1959
- >=1960 and <=1969
- >=1970 and <=1979
- >=1980 and <=1989
- >=1990 and <=1999
- >=2000

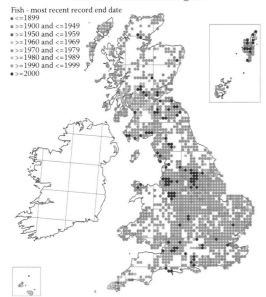

The Three-Spined Stickleback (*Gasterosteus aculeatus*)

### *Pungitius pungitius* Fsh_313

Fish - most recent record end date
- <=1899
- >=1900 and <=1949
- >=1950 and <=1959
- >=1960 and <=1969
- >=1970 and <=1979
- >=1980 and <=1989
- >=1990 and <=1999
- >=2000

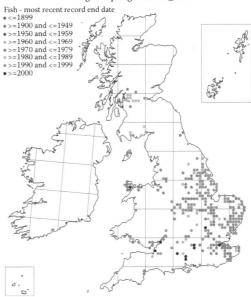

The Ten-Spined or Nine-Spined Stickleback (*Pungitius pungitius*)

### *Cottus gobio* Fsh_208

Fish - most recent record end date
- <=1899
- >=1900 and <=1949
- >=1950 and <=1959
- >=1960 and <=1969
- >=1970 and <=1979
- >=1980 and <=1989
- >=1990 and <=1999
- >=2000

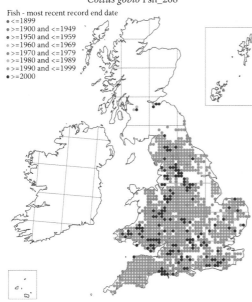

The Bullhead (*Cottus gobio*)

### *Barbatula barbatula* Fsh_166

Fish - most recent record end date
- <=1899
- >=1900 and <=1949
- >=1950 and <=1959
- >=1960 and <=1969
- >=1970 and <=1979
- >=1980 and <=1989
- >=1990 and <=1999
- >=2000

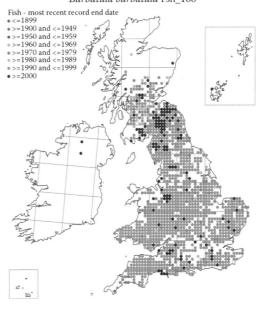

The Stone Loach (*Barbatula barbatula*)

*Cobitis taenia* Fsh_199

Fish - most recent record end date
- <=1899
- >=1900 and <=1949
- >=1950 and <=1959
- >=1960 and <=1969
- >=1970 and <=1979
- >=1980 and <=1989
- >=1990 and <=1999
- >=2000

The Spined Loach (*Cobitis taenia*)

## A2.7 THE ELUSIVE EUROPEAN EEL

*Anguilla anguilla* Fsh_157

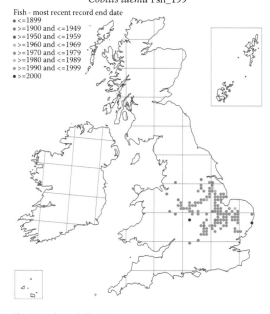

Fish - most recent record end date
- <=1899
- >=1900 and <=1949
- >=1950 and <=1959
- >=1960 and <=1969
- >=1970 and <=1979
- >=1980 and <=1989
- >=1990 and <=1999
- >=2000

The European Eel (*Anguilla anguilla*)

# A2.8 BRITAIN'S ANCIENT LAMPREYS

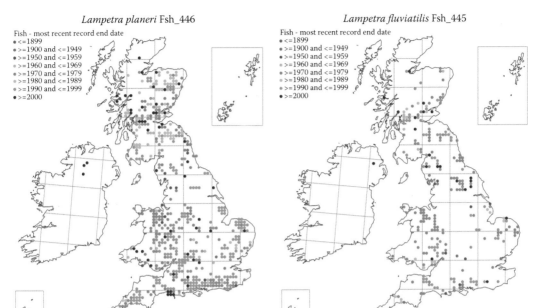

*Lampetra planeri* Fsh_446

Fish - most recent record end date
- <=1899
- >=1900 and <=1949
- >=1950 and <=1959
- >=1960 and <=1969
- >=1970 and <=1979
- >=1980 and <=1989
- >=1990 and <=1999
- >=2000

The Brook Lamprey (*Lampetra planeri*)

*Lampetra fluviatilis* Fsh_445

Fish - most recent record end date
- <=1899
- >=1900 and <=1949
- >=1950 and <=1959
- >=1960 and <=1969
- >=1970 and <=1979
- >=1980 and <=1989
- >=1990 and <=1999
- >=2000

The River Lamprey (*Lampetra fluviatilis*)

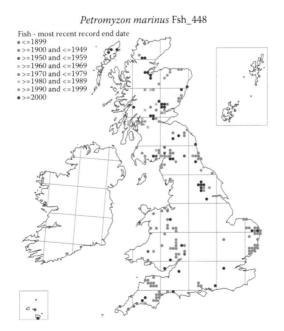

*Petromyzon marinus* Fsh_448

Fish - most recent record end date
- <=1899
- >=1900 and <=1949
- >=1950 and <=1959
- >=1960 and <=1969
- >=1970 and <=1979
- >=1980 and <=1989
- >=1990 and <=1999
- >=2000

The Sea Lamprey (*Petromyzon marinus*)

## A2.9  THE BRITISH SHADS

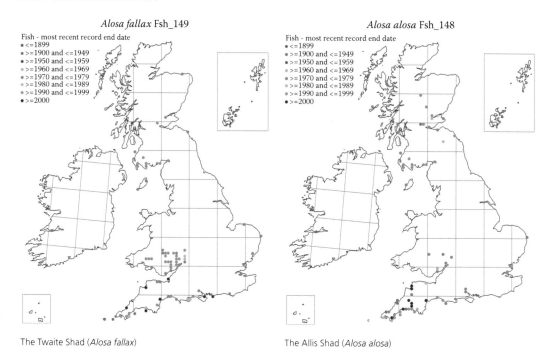

*Alosa fallax* Fsh_149

Fish - most recent record end date
- ● <=1899
- ● >=1900 and <=1949
- ● >=1950 and <=1959
- ● >=1960 and <=1969
- ● >=1970 and <=1979
- ● >=1980 and <=1989
- ● >=1990 and <=1999
- ● >=2000

*Alosa alosa* Fsh_148

Fish - most recent record end date
- ● <=1899
- ● >=1900 and <=1949
- ● >=1950 and <=1959
- ● >=1960 and <=1969
- ● >=1970 and <=1979
- ● >=1980 and <=1989
- ● >=1990 and <=1999
- ● >=2000

The Twaite Shad (*Alosa fallax*)

The Allis Shad (*Alosa alosa*)

## A2.10  THE NOT-SO-COMMON STURGEON

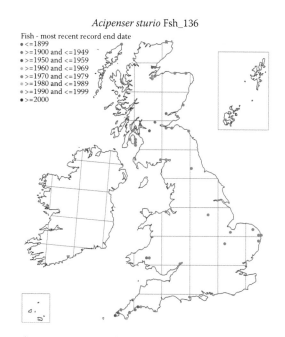

*Acipenser sturio* Fsh_136

Fish - most recent record end date
- ● <=1899
- ● >=1900 and <=1949
- ● >=1950 and <=1959
- ● >=1960 and <=1969
- ● >=1970 and <=1979
- ● >=1980 and <=1989
- ● >=1990 and <=1999
- ● >=2000

The Common Sturgeon (*Acipenser sturio*)

## A2.11 THE BURBOT, NOW EXTINCT IN BRITAIN

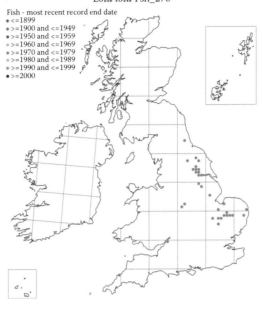

*Lota lota* Fsh_270

Fish - most recent record end date
- ● <=1899
- ● >=1900 and <=1949
- ● >=1950 and <=1959
- ● >=1960 and <=1969
- ● >=1970 and <=1979
- ● >=1980 and <=1989
- ● >=1990 and <=1999
- ● >=2000

The Burbot (*Lota lota*)

## A2.12 ALIEN CATFISHES AND PUMPKINSEED

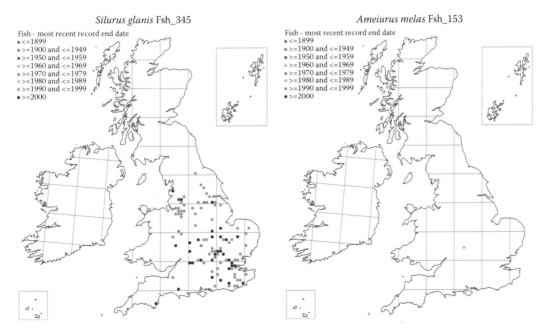

*Silurus glanis* Fsh_345

Fish - most recent record end date
- ● <=1899
- ● >=1900 and <=1949
- ● >=1950 and <=1959
- ● >=1960 and <=1969
- ● >=1970 and <=1979
- ● >=1980 and <=1989
- ● >=1990 and <=1999
- ● >=2000

*Ameiurus melas* Fsh_153

Fish - most recent record end date
- ● <=1899
- ● >=1900 and <=1949
- ● >=1950 and <=1959
- ● >=1960 and <=1969
- ● >=1970 and <=1979
- ● >=1980 and <=1989
- ● >=1990 and <=1999
- ● >=2000

The Wels Catfish (*Siluris glanis*)

The Black Bullhead (*Ameiurus melas*)

## *Lepomis gibbosus* Fsh_250

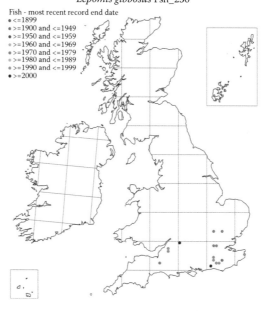

Fish - most recent record end date
- <=1899
- >=1900 and <=1949
- >=1950 and <=1959
- >=1960 and <=1969
- >=1970 and <=1979
- >=1980 and <=1989
- >=1990 and <=1999
- >=2000

The Pumpkinseed (*Lepomis gibbosus*)

## A2.13 MARINE INTRUDERS

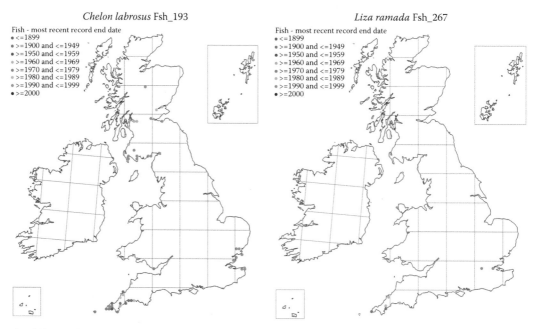

### *Chelon labrosus* Fsh_193

Fish - most recent record end date
- <=1899
- >=1900 and <=1949
- >=1950 and <=1959
- >=1960 and <=1969
- >=1970 and <=1979
- >=1980 and <=1989
- >=1990 and <=1999
- >=2000

The Thick-Lipped Grey Mullet (*Chelon labrosus*)

### *Liza ramada* Fsh_267

Fish - most recent record end date
- <=1899
- >=1900 and <=1949
- >=1950 and <=1959
- >=1960 and <=1969
- >=1970 and <=1979
- >=1980 and <=1989
- >=1990 and <=1999
- >=2000

The Thin-Lipped Grey Mullet (*Liza ramada*)

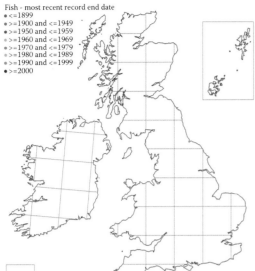

### Liza aurata Fsh_266

Fish - most recent record end date
- ≤=1899
- >=1900 and <=1949
- >=1950 and <=1959
- >=1960 and <=1969
- >=1970 and <=1979
- >=1980 and <=1989
- >=1990 and <=1999
- >=2000

The Golden-Grey Mullet (*Liza aurata*)

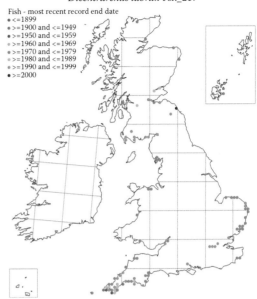

### Dicentrarchus labrax Fsh_217

Fish - most recent record end date
- ≤=1899
- >=1900 and <=1949
- >=1950 and <=1959
- >=1960 and <=1969
- >=1970 and <=1979
- >=1980 and <=1989
- >=1990 and <=1999
- >=2000

The European Seabass (*Dicentrarchus labrax*)

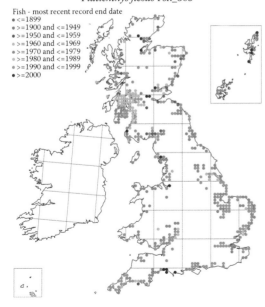

### Platichthys flesus Fsh_305

Fish - most recent record end date
- ≤=1899
- >=1900 and <=1949
- >=1950 and <=1959
- >=1960 and <=1969
- >=1970 and <=1979
- >=1980 and <=1989
- >=1990 and <=1999
- >=2000

The Flounder (*Platichthys flesus*)

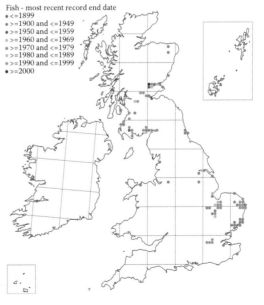

### Osmerus eperlanus Fsh_294

Fish - most recent record end date
- ≤=1899
- >=1900 and <=1949
- >=1950 and <=1959
- >=1960 and <=1969
- >=1970 and <=1979
- >=1980 and <=1989
- >=1990 and <=1999
- >=2000

The Smelt (*Osmerus eperlanus*)

The Sand Smelt (*Atherina presbyter*): No distribution map is available for this species.

# Appendix 3: The tendency of British freshwater leuciscine fishes to hybridise

| | Roach | Rudd | Silver bream | Common bream | Chub | Dace | Bleak |
|---|---|---|---|---|---|---|---|
| **Roach** | | | | | | | |
| **Rudd** | Yes | | | | | | |
| **Silver bream** | Yes | Yes | | | | | |
| **Common bream** | Yes | Yes | Yes | | | | |
| **Chub** | Rarely | No | No | No | | | |
| **Dace** | No | No | No | No | No | | |
| **Bleak** | Rarely | No | Rarely | Rarely | No | No | |
| **Orfe** | No | No | No | No | No | Rarely | No |

# Appendix 4: Key identification features of British freshwater cyprinid fishes

| | Native or introduced | Barbels | Scales | Body shape | Scales along lateral line | Spines/ rays in dorsal fin | Spines/ rays in anal fin |
|---|---|---|---|---|---|---|---|
| **Roach** (*Rutilus rutilus*) | Native | Absent | Conspicuous | Deep and laterally compressed | 42–45 | III/9–11 | III/9–11 |
| **Rudd** (*Scardinius erythrophthalmus*) | Native | Absent | Conspicuous | Deep and laterally compressed | 40–45 | III/8–9 | III/10–11 |
| **Chub** (*Squalius cephalus*) | Native | Absent | Conspicuous | Streamlined and round in cross section | 44–46 | III/8–9 | III/7–9 |
| **Dace** (*Leuciscus leuciscus*) | Native | Absent | Conspicuous | Streamlined and round in cross section | 48–51 | III/7 | III/8 |
| **Orfe** (*Leuciscus idus*) | Introduced, locally naturalised | Absent | Conspicuous | Streamlined and round in cross section | 56–61 | III/8 | III/9–10 |
| **Common bream** (*Abramis brama*) | Native | Absent | Conspicuous | Deep and laterally compressed | 51–60 | III/9 | III/24–30 |
| **Silver bream** (*Blicca bjoerkna*) | Native | Absent | Conspicuous | Deep and laterally compressed | 44–48 | III/8–9 | III/21–23 |
| **Bleak** (*Alburnus alburnus*) | Native | Absent | Conspicuous | Streamlined and laterally compressed | 48–55 | III/8–9 | III/16–20 |
| **Common carp** (*Cyprinus carpio*) | Introduced periodically since 15th century, widely naturalised | Two pairs | Conspicuous (or rarely absent) | Deep and laterally compressed | 35–40 | III–IV/17–22 | II–III/5 |

*(Continued)*

| | Native or introduced | Barbels | Scales | Body shape | Scales along lateral line | Spines/rays in dorsal fin | Spines/rays in anal fin |
|---|---|---|---|---|---|---|---|
| **Crucian carp** (*Carassius carassius*) | Native | Absent | Conspicuous | Deep and laterally compressed | 32–25 | III–IV/14–21 | III/6–8 |
| **Goldfish** (*Carassius auratus*) | Introduced, widely naturalised | Absent | Conspicuous | Deep and laterally compressed | 28–33 | III–IV/15–19 | II–III/5–6 |
| **Grass carp** (*Ctenopharyngodon idella*) | Introduced, non-breeding, locally established | Absent | Conspicuous | Streamlined and round in cross section | 40–42 | III/7–8 | III/7–11 |
| **Barbel** (*Barbus barbus*) | Native | Two pairs | Conspicuous | Streamlined and flattened underneath | 55–65 | III/7–9 | III/5 |
| **Gudgeon** (*Gobio gobio*) | Native | One pair | Conspicuous | Streamlined and flattened underneath | 38–44 | III/5–7 | III/6–7 |
| **Minnow** (*Phoxinus phoxinus*) | Native | Absent | Very small, apparently scaleless | Streamlined and round in cross section | Lateral line incomplete | III/7 | III/6–7 |
| **Tench** (*Tinca tinca*) | Native | One pair | Small | Oval and round in cross section | Many small scales | III/8 | III/6–8 |
| **Bitterling** (*Rhodeus amarus*) | Introduced, locally naturalised | Absent | Conspicuous | Deep and laterally compressed | 34–38 | III/9–10 | III/8–9 |
| **Topmouth gudgeon** (*Pseudorasbora parva*) | Introduced, invasive, locally established | Absent | Conspicuous | Streamlined and round in cross section | 34–38 | III/7 | III/6 |
| **Sunbleak** (*Leucaspius delineatus*) | Introduced, invasive, locally established | Absent | Conspicuous | Streamlined and round in cross section | Incomplete lateral line with about 8–12 pored scales | II–III/7–9 | III/10–13 |

# Appendix 5: Freshwater fishes banned under UK legislation

**TEMPERATE FRESHWATER FISH FOR WHICH TRADE AND KEEPING IS RESTRICTED IN ENGLAND AND WALES**

*The following fish species are scheduled under the Import of Live Fish (England and Wales) Act 1980*

- Blue bream (*Abramis ballerus*)
- Sterlets and sturgeons (species of the genera *Acipenser, Huso, Scaphirhynchus, Pseudoscaphirhynchus* and their hybrids)
- Schneider (*Alburnoides bipunctatus*)
- Rock bass (*Ambloplites rupestris*)
- Coldwater ameiurid catfishes (*Ameiurus* spp.), including the bullhead (*Ameirus nebulosus*)
- Asp (*Aspius aspius*)
- Danubian bleak (*Chalcaburnus chalcoides*)
- Nase (*Chrondrostoma nasus*)
- Toxostome or French nase (*Chrondrostoma toxostoma*)
- Grass carp (*Ctenopharyngodon idella*)
- Silver carp (*Hypophthalmichthys molitrix*)
- Coldwater ictalurid catfishes (*Ictalurus* spp.), including the channel catfish (*Ictalurus punctatus*)
- Blageon (*Leuciscus souffia*)
- Burbot (*Lota lota*)
- Large-mouthed bass (*Micropterus salmoides*)
- Black or snail-eating carp (*Mylopharyngodon piceus*)
- Rainbow trout or steelhead (*Oncorhynchus mykiss*)
- Pacific trout (*Oncorhynchus* spp.)
- Paddlefishes (*Polyodon spathula* and *Psephurus gladius*)
- Clicker barb or topmouth gudgeon (*Pseudorasbora parva*)
- European bitterling (*Rhodeus amarus*)
- Non-anadromous, landlocked salmon (*Salmo salar*)
- Coldwater silurid catfishes (*Silurus* spp.), including the wels catfish (*Silurus glanis*)
- Zander (*Stizostedion* spp.)
- Vimba (*Vimba vimba*)

*New species included under the Prohibition of Keeping or Release of Live Fish (Specified Species) Order 1998*

- Barbel species (*Barbus* spp.), excluding the native (*Barbus barbus*)
- Common white sucker (*Catostomus commersoni*)
- Northern snakehead (*Channa argus*)
- Whitefishes (*Coregonus* spp.), excluding the native European whitefish *C. lavaretus* and vendace *C. albula*

- Blue sucker (*Cycleptus elongatus*)
- Red shiner or rainbow dace (*Cyprinella* (*Notropsis*) *lutrensis*)
- Pikes (*Esox* spp.), excluding the native *E. lucius*
- Danubian salmon or taimen (*Hucho* spp.)
- *Lepomis* species (pumpkinseeds, sunfish, sunbass, crappies, bluegills and other)
- Motherless minnow or sunbleak (*Leucaspius delineatus*)
- Weather loach (*Misgurnus fossilis*)
- *Morone* species (striped bass, white bass and morone hybrids)
- Chinese sailfin sucker (*Myxocyprinus asiaticus*)
- Perch species (*Perca* spp.), excluding the native perch (*P. fluviatilis*)
- Northern red-belly dace (*Phoxinus* (*Chrosomus*) *eos*)
- Southern red-belly dace (*Phoxinus* (*Chrosomus*) *erythrogaster*)
- Rosy-red minnow or fathead minnow (*Pimephales promelas*)
- Blacknose dace (*Rhinichthys atratulus*)
- Marbled trout (*Salmo marmoratus*)
- Charr species (*Salvelinus* spp.), including the American brook trout, but excluding the native Arctic charr (*Salvelinus alpinus*)
- European mudminnow (*Umbra krameri*)
- Eastern mudminnow (*Umbra pygmaea*)
- Pale chub (*Zacco platypus*)

# Appendix 6: Fish species scheduled for conservation action

This appendix describes key international agreements and legislation, both European and domestic, scheduling British freshwater fish species for special conservation action.

## A6.1 THE IUCN RED LIST

The International Union for Conservation of Nature (IUCN) Red List of Threatened Species, also known as The IUCN Red List or Red Data List, is a comprehensive and regularly updated inventory of the global conservation status of species of plants and animals. Both global and regional lists are maintained, classifying species according to extinction risk based on established criteria. These criteria include the rate of decline, population size, area of geographic distribution and degree of population and distribution fragmentation. The purpose of the Red List is to communicate conservation issues to the public and policy makers. Species are categorized as either Extinct (EX), Extinct in the Wild (EW), Critically Endangered (CR), Endangered (EN), Vulnerable (VU), Near Threatened (NT), Least Concern (LC), Data Deficient (DD) or Not Evaluated (NE). The term 'threatened' is a grouping of the Critically Endangered, Endangered and Vulnerable categories. Many British freshwater fish species are categorised as Least Concern (LC). However, those identified at higher risk of extinction on the Red List at the time of writing are noted below.

The Red List reflects the global range of the listed species, creating some disconnections with conservation status in Britain. For example, listed as Least Concern (LC) are vendace, twaite shad and allis shad, all highly localised and vulnerable in Britain, whilst the crucian carp is widely threatened by hybridisation and loss of genetically intact populations. Topmouth gudgeon and sunbleak too are listed as Least Concern (LC), yet as alien invasive species in British waters, they represent a significant threat to native ecosystems. The burbot too is listed as of Least Concern (LC), yet is extinct in Britain, and six out of seven of the species listed as Not Evaluated (NE) are alien introductions here. Nevertheless, applied within an appropriate local context, the Red List is a useful and consensual tool for prioritising conservation action relating the conservation of species across their global ranges.

| Red list category | Species | |
|---|---|---|
| Least Concern (LC) | • Roach | • Ruffe |
| | • Rudd | • Zander |
| | • Chub | • Bullhead |
| | • Dace | • Three-spined stickleback |
| | • Common bream | • Ten-spined stickleback |
| | • Silver bream | • Stone loach |
| | • Bleak | • Spined loach |
| | • Orfe | • Allis shad |
| | • Barbel | • Twaite shad |
| | • Crucian carp | • Brook lamprey |
| | • Tench | • River lamprey |
| | • Gudgeon | • Sea lamprey |
| | • Minnow | • Burbot |
| | • Bitterling | • Wels catfish |
| | • Sunbleak | • European seabass |
| | • Atlantic salmon | • Flounder |
| | • Brown/Sea trout | • Smelt |
| | • Grayling | • Sand smelt |
| | • Arctic charr | • Thick-lipped grey mullet |
| | • Vendace | • Thin-lipped grey mullet |
| | • Pike | • Golden grey mullet |
| | • Perch | |
| Vulnerable (VU) | • Common carp (wild form) | |
| Critically Endangered (CR) | • European eel | |
| | • Common sturgeon | |
| | • Gwyniad (recognised on the Red List as a separate species, though more generally as a local population of the European whitefish *Corgonus lavaretus*) | |
| Extinct (EX) | • Houting | |
| Not Evaluated (NE) | • Sand smelt | |
| | • Rainbow trout | |
| | • Goldfish | |
| | • Grass carp | |
| | • Topmouth gudgeon | |
| | • Pumpkinseed sunfish | |
| | • Black bullhead | |

Appendix 6: Fish species scheduled for conservation action

## A6.2 THE BERN CONVENTION

The Bern Convention on the Conservation of European Wildlife and Natural Habitats 1979, generally referred to as the Bern Convention (or Berne Convention), came into force in 1982. Its aims are to conserve wild flora and fauna and their natural habitats, promote cooperation between states, monitor and control endangered and vulnerable species and help with the provision of assistance concerning legal and scientific issues. The Bern Convention led to the creation in 1998 of the Emerald Network of Areas of Special Conservation Interest (ASCIs) throughout the territory of the parties to the Convention, which operates alongside the EU Natura 2000 programme. It also provides for the monitoring and control of endangered species and the provision of assistance concerning legal and scientific issues. Four appendices, each regularly updated by Expert Groups, set out particular species for protection. These are listed below, together with affected species.

The Bern Convention reflects the conservation status of species across their European range, so it also has to be interpreted and applied within a national context. For example, included in Appendix III are sunbleak, wels catfish and bitterling, all three of them not native to Britain and the sunbleak proving to be a problematic invasive species.

| Bern Convention category |
|---|
| • Listed freshwater fish species |
| Appendix I addresses strictly protected flora species and therefore does not directly concern fish. |
| Appendix II lists strictly protected fauna species. Scheduled British freshwater fish species include: |
| • Common sturgeon |
| Appendix III is concerned with species that are regulated but the exploitation of which is controlled in accordance with the Directive. Scheduled British freshwater fish species include: |

- River lamprey
- Brook lamprey
- Sea lamprey
- Allis shad
- Twaite shad
- Grayling
- Atlantic salmon
- Spined loach
- Sunbleak
- Wels catfish
- Bitterling

Appendix IV prohibited means and methods of killing, capture and other forms of exploitation, including:

- Explosives
- Firearms
- Poisons
- Anaesthetics
- Electricity with alternating current
- Artificial light sources

## A6.3 THE EUROPEAN UNION HABITATS DIRECTIVE

The European Union (EU) Habitats Directive (Council Directive 92/43/EEC on the Conservation of Natural Habitats and of Wild Fauna and Flora) was adopted in 1992 as one of the cornerstones, together with the EU Birds Directive, of Europe's nature conservation policy. The Habitats Directive, including subsequent amendments, comprises two pillars: the Natura 2000 network of protected sites and a strict system of species protection. One thousand animal and plant species and over two hundred 'habitat types' of European importance are covered by the Habitats Directive. The Habitats Directive is also the means by which the EU meets its obligations under the Bern Convention. The Directive thereby promotes the maintenance of biodiversity by requiring member states to take measures to maintain or restore natural habitats and wild species listed in the Annexes to the Directive at a favourable conservation status through robust protection measures that also take account of economic, social and cultural requirements, as well as regional and local characteristics. These Annexes and their associated species are listed below.

| EU Habitats Directive Annex |
|---|
| • Listed freshwater fish species |
| Annex I: Natural habitat types of community interest whose conservation requires the designation of Special Areas of Conservation (SAC). This comprises a list of habitat types, some of which are relevant to various British freshwater fish species. |
| Annex II: Animal and plant species of community interest whose conservation requires the designation of Special Areas of Conservation (SAC). These include (with some regional exclusions): <br>• River lamprey <br>• Brook lamprey <br>• Sea lamprey <br>• Common sturgeon <br>• Allis shad <br>• Twaite shad <br>• Atlantic salmon <br>• Bitterling (an introduced species in the UK) <br>• Spined loach <br>• Bullhead |
| Annex III: Criteria for selecting sites eligible for identification as sites of community importance and designation as Special Areas of Conservation (SAC). This includes site selection based on, amongst other criteria, the needs of species listed in Annex II. |
| Annex IV: Animal and plant species of community interest in need of strict protection. This includes the following British fishes (with regional exemptions for some species): <br>• Common sturgeon <br>• Houting (the IUCN has classified this species as Extinct) |

(Continued)

Annex V: Animal and plant species of community interest whose taking in the wild and exploitation may be subject to management measures. This Annex includes:

- River lamprey
- Allis shad
- Twaite shad
- Grayling
- All species of *Coregonus* except houting, thus including:
  - Vendace
  - European whitefish (including gwyniad, powan, pollan and schelly)
- Atlantic salmon
- Barbel

Annex VI: Prohibited methods and means of capture and killing and modes of transport, including:

- Poison
- Explosives

In the UK, the Directive has been transposed into law by The Conservation (Natural Habitats, &c.) Regulations 1994 and The Conservation of Habitats and Species Regulations 2010 consolidating various amendments. In Scotland, the Habitats Directive is transposed through a combination of the Habitats Regulations 2010 (in relation to reserved matters) and the UK's 1994 Regulations. The Conservation (Natural Habitats, &c.) Regulations (Northern Ireland) 1995 (as amended) transpose the Habitats Directive in relation to Northern Ireland. The Offshore Marine Conservation (Natural Habitats etc.) Regulations 2007 (as amended) address EU Habitats Directive requirements for UK offshore waters (from twelve nautical miles from the coast out to two hundred nautical miles or to the limit of the UK Continental Shelf Designated Area).

### A6.4 THE WILDLIFE AND COUNTRYSIDE ACT 1981

The Wildlife and Countryside Act 1981 (as subsequently amended) is the means by which the UK implemented the EU Birds Directive (Directive 2009/147/EC on the Conservation of Wild Birds) but also consolidates various other forms of protection of wildlife. The Act gives protection to native species (especially those under threat), controls the release of non-native species, enhances the protection of Sites of Special Scientific Interest (SSSIs) and builds upon the rights of way rules in the National Parks and Access to the Countryside Act 1949. It comprises four parts and is supported by seventeen schedules, the relevant ones and associated species listed below.

- Listed freshwater fish species

Schedule 5: Animals which are protected (in England and Wales)

- Allis shad
- Twaite shad
- Common sturgeon
- Vendace
- European whitefish

Schedule 9: Animals and plants to which section 14 applies (Section 14 relates to the 'Introduction of new species etc'.)

- Bitterling
- Pumpkinseed sunfish
- Wels catfish
- Zander

Schedule 10: Amendment of the Endangered Species (Import and Export) Act 1976. This affects a number of fish and other species prohibited from import into the UK

## A6.5 THE UK BIODIVERSITY ACTION PLAN

The UK Biodiversity Action Plan (UK BAP), published in 1994, is the UK Government's response to the Convention on Biological Diversity (CBD) to which the UK became a signatory at the 1992 Rio de Janeiro 'Earth Summit'. The CBD called for the development and enforcement of national strategies and associated action plans to identify, conserve and protect existing biological diversity and to enhance it wherever possible. British freshwater fish species listed under the UK BAP include:

- Common sturgeon (not in Wales)
- Allis shad
- Twaite shad
- European eel
- Spined loach (not in Scotland, Wales or Northern Ireland)
- Vendace (not in Wales or Northern Ireland)
- European whitefish (including powan, pollan, gwyniad and schelly) (not pollan in Northern Ireland)
- River amprey
- Burbot (though noted as extinct in England, Wales, Scotland and Northern Ireland)
- Smelt (sparling)
- Sea lamprey
- Atlantic salmon
- Brown trout
- Arctic charr (not in Northern Ireland)

# Index

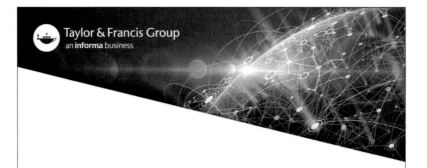